Adsorption at Interfaces

Adsorption at Interfaces

K. L. Mittal, *Editor*

Papers from a symposium honoring

Robert D. Vold and Marjorie J. Vold

sponsored by the Division of

Colloid and Surface Chemistry

at the 167th Meeting of the

American Chemical Society

Los Angeles, Calif.,

April 2–5, 1974.

ACS SYMPOSIUM SERIES 8

AMERICAN CHEMICAL SOCIETY

WASHINGTON, D. C.   1975

Library of Congress CIP Data

Adsorption at interfaces.
  (ACS symposium series; no. 8)

  Companion vol. to Colloidal dispersions and micellar
behavior.

  Includes bibliographical references and index.

  1. Adsorption—Congresses.
  I. Mittal, K. L., 1945- ed. II. American Chemical
Society. Division of Colloids and Surface Chemistry.
III. American Chemical Society. IV. Series: American
Chemical Society. ACS symposium series; no. 8.

QD547.A37   541'.3453   74-32040
ISBN 8412-0249-4   ACSMC8 8 1-290 (1975)

ACS Symposium Series

Robert F. Gould, *Series Editor*

FOREWORD

The ACS SYMPOSIUM SERIES was founded in 1974 to provide a medium for publishing symposia quickly in book form. The format of the SERIES parallels that of its predecessor, ADVANCES IN CHEMISTRY SERIES, except that in order to save time the papers are not typeset but are reproduced as they are submitted by the authors in camera-ready form. As a further means of saving time, the papers are not edited or reviewed except by the symposium chairman, who becomes editor of the book. Papers published in the ACS SYMPOSIUM SERIES are original contributions not published elsewhere in whole or major part and include reports of research as well as reviews since symposia may embrace both types of presentation.

CONTENTS

PREFACE

Professors R. D. Vold and M. J. Vold retired in June 1974 after more than 30 yrs at the University of Southern California. The annual meeting of the American Chemical Society in April in near-by Los Angeles provided an ideal opportunity to pay tribute to these two outstanding workers in colloid science by holding a symposium in their honor.

When the idea for this symposium was broached to me, I accepted it without hesitation as I knew of their global popularity and knew that such an event would elicit an enthusiastic response both from their former students and fellow colloid scientists. The response exceeded all expectations as 57 papers covering a wide variety of interfacial and colloidal phenomena by 98 authors from 14 countries were included in the program, and two papers were considered later on. Such overwhelming response certainly testifies to their popularity as well as versatility in this research field. In fact, this symposium turned out to be the second largest one at the meeting. In consideration of the contents of this symposium, it might be more appropriate to name it "International Symposium on Interfacial and Colloid Phenomena honoring Professors R. D. and M. J. Vold."

This volume of 20 papers documents part of the proceedings of the symposium. The other part is contained in a companion volume of 24 papers, entitled "Colloidal Dispersions and Micellar Behavior." The papers in this volume deal with the adsorption of a variety of adsorbates on an array of substrates. Thermodynamics of adsorption, insoluble monolayers, adsorption at low energy solids, adsorption at colloidal particles, adsorption of polymers, and other aspects of adsorption are covered.

Adsorption plays an important role in many technological, industrial, natural, and biological processes. Adsorption is truly an interdisciplinary field as is evidenced by the vast amount of literature being published from diverse laboratories. Adsorption at interfaces ranges from the adsorption of simple molecules (for example, gases) on bulk solid substrates to the adsorption of polymeric materials on colloidal particles. Innumerable adsorption applications have produced a proliferation of literature on this topic. With the availability of sophisticated instrumentation, a tremendous progress has been made in understanding the nature of the adsorbed species and the absorbate–adsorbent interactions, but the subject of adsorption is still pregnant with many challenging problems whose solutions will open new vistas in many basic and applied areas.

Colloid chemistry, more generally colloid science, has been aptly described by the Volds in their book "Colloid Chemistry" (Reinhold, 1964) as "the science of large molecules, small particles, and surfaces." So obviously, adsorption is an integral part of the study of colloids, and many of their characteristics are attributable to the adsorption at colloidal particles.

The Volds have been active in a spectrum of research activities. He made his research debut in the study of the solubility of organic compounds in aqueous and nonaqueous media while her first paper was on the mechanism of substitution reactions. After this Robert worked in industry where he became acutely aware of the inadequacies of the theories of simple systems to practical industrial problems and decided to study the intricacies of colloidal systems. Subsequently, Marjorie became very interested in studying colloids. They actively pursued the phase rule studies of association colloids (soaps, greases, etc.) in aqueous and nonaqueous media using surface chemical, electron microscopic, x-ray diffraction, and thermal analytical (DTA) techniques; adsorption at various interfaces; stability of colloidal dispersions; and calculation of van der Waals forces. More recently, their research interests have included understanding the factors influencing emulsion stability using ultracentrifuge; use of computer in floc formation and calculation of the dimensions of coiling type polymers; dispersions of carbon black; phase behavior of lithium stearate greases; theories of colloidal stability in nonaqueous media; and the hydration of biopolymers like DNA.

Obviously, the Volds' research activities have run the gamut from less glamorous colloidal systems like greases to the more fashionable biopolymers. Their work on the phase behavior and properties of nonaqueous soap systems had a significant impact in the petroleum industry (cf. NLGI Spokesman **18**, 168 (1964)). Their research investigations have culminated in 136 scientific and technical publications. They have also written the book mentioned above. This small paperback is an extremely good exposition of the principles and the methods of study of colloidal systems. Owing to its popularity and utility it has been translated into Japanese.

Apart from their research contributions, the Volds have rendered a great service to colloid science by popularizing it on a global basis, and they were very instrumental, along with Professors Adamson, Mysels, and Simha, in establishing an internationally acclaimed center for surface and colloid chemistry in the Chemistry Department at the University of Southern California. Robert Vold organized the Summer Conferences on Colloid Chemistry (1961–1964) at U.S.C. sponsored by the National Science Foundation.

The Volds are very meticulous researchers and highly stimulating teachers. I have found them very adept at inculcating good research habits in the minds of neophytes in research. No student can get away with sloppy record keeping. Furthermore, they know how to bring out the best in a student.

Robert D. Vold received his A.B. and M.S. degrees at the University of Nebraska in 1931 and 1932, respectively, and Ph.D. degree from the University of California at Berkeley in 1935. During 1935–1937 he was a research chemist with Proctor and Gamble in Cincinnati. He was a postdoctoral fellow with Professor J. W. McBain at Stanford University from 1937 to 1941. Since 1941 he has been engaged in teaching and research at the University of Southern California. In 1953–1954 he held a Fulbright Senior Research Fellowship with Professor Overbeek at the University of Utrecht, The Netherlands. In 1955–1957 he served as Visiting Professor of Physical Chemistry at the Indian Institute of Science, Bangalore, India, where he introduced new fields of research and helped to establish a Ph.D. program. In 1965 he served as a consultant for the Summer Institute at Jadavpur University, Calcutta, India, which is designed to improve the teaching of chemistry in Indian colleges. He was a member of the advisory board of the Journal of Colloid Science from its inception in 1946 until 1960.

Among the various offices he has held in National Associations include Chairman, Southern California Section, ACS, 1967; Committee on Colloid and Surface Chemistry of the National Academy of Sciences, 1964–1967; Chairman, California Association of Chemistry Teachers, 1961; National Colloid Symposium Committee, 1948–1953; and Chairman, Division of Colloid Chemistry, ACS, 1947–1948. He was awarded the Tolman medal of the Southern California Section of the American Chemical Society for his research contributions and service to the profession in 1970.

Marjorie J. Vold received her B.S. in 1934 and Ph.D. in 1936 from the University of California at Berkeley at the unusually young age of 23. She was University Medallist (Class Valedictorian) at U.C. Berkeley. After brief experience as a lecturer at the University of Cincinnati and the University of Southern California, she was a research chemist with the Union Oil Co. (1942–1946). Since then she has held various faculty appointments in the Department of Chemistry of the University of Southern California, with the title of Adjunct Professor for the last 14 yrs. She was awarded a Guggenheim Fellowship in 1953 which was taken at the University of Utrecht, The Netherlands. From 1967 to 1970 she served as a member of the Advisory Board of the Journal of Colloid and Interface Science.

She was awarded the Garvan Medal of the American Chemical Society in 1967. Among many other awards and honors include Los Angeles Times "Woman of the Year," 1966; National Lubricating Grease Institute Authors Award for the "Best Paper" presented at the Annual Meeting (in 1967), 1968; and she is listed among the 100 Outstanding Women of the U.S. by the Women's Home Companion 1969. Unfortunately, Mrs. Vold was struck with multiple sclerosis in the fall of 1958 and has been confined to a wheel chair most of the time since 1960. In spite of her poor health and this terrible handicap, she has shown admirable stamina to deliver advanced colloid chemistry lectures continuously for two hours.

On November 19, 1967 an article entitled "Science: A Tie That Binds" appeared in Los Angeles Herald Examiner which presented a few glimpses of the scientific and social lives of the Volds. A special love for science has been maintained in the Vold clan for almost a century. Individually speaking, the Volds are different in many respects. Mrs. Vold has put it succinctly, "Robert has a sound almost conservative judgment while I tend to go off half-cocked. We complement each other in our work and our lives."

Although the Volds are retiring from active duty, they have no intention of relinquishing their interest in colloid science—a discipline they have cherished for about 40 yrs—as they plan to write a text book on the subject.

May we join together on this occasion to wish them a very healthy and enjoyable retirement in San Diego.

Acknowledgments: First, I am grateful to the Division of Colloid and Surface Chemistry for sponsoring this event. I am greatly indebted to the management of the IBM Corp., both at San Jose and at Poughkeepsie, for permitting me to organize this symposium and edit the volumes. Special thanks are due to my manager, E. L. Joba, for his patience and understanding. The secretarial assistance of Carol Smith is gratefully acknowledged. Special thanks are also due to M. J. Dvorocsik and Elizabeth M. Ragnone for helping to prepare this volume for publication. The able guidance and ready and willing help of Paul Becher and K. J. Mysels is deeply appreciated. The reviewers should be thanked for their many valuable comments on the manuscripts. I am thankful to my wife, Usha, for helping with the correspondence, proofreading, and above all for tolerating, without complaint, the frequent privations of an editor's wife. It would be remiss on my part if I failed to acknowledge the enthusiasm and cooperation of all the participants, especially the delegates from overseas countries.

Poughkeepsie, N.Y., November 19, 1974   K. L. MITTAL

Adsorption at Interfaces

Thermodynamics of Adsorption and Interparticle Forces

D. H. EVERETT
School of Chemistry, University of Bristol, Bristol BS8 1TS, England
C. J. RADKE
Chemical Engineering Department, Pennsylvania State University, University Park, Penn. 16802

Introduction

An understanding of the properties of colloidal systems depends critically upon a knowledge of the factors which determine the forces between small particles in a fluid medium. Many systems of practical importance are dispersions in aqueous electrolyte solutions and it was natural that the earlier theories of colloid stability were concerned with the way in which electrostatic forces, arising from ions in the solution and the electrical state of the surface, combine with dispersion forces between the particles to determine the thermodynamically stable (or metastable) state of the system. With the development of systems (often in non-aqueous media) stabilised by adsorbed molecules, attention was directed towards the effect of adsorbed species on interparticle forces. A number of possible contributions to these effects were recognised but in earlier treatments they were included as separate additive contributions to the interaction energy. The phenomena associated with adhesion between particles in powders have been looked upon as unrelated problems and no quantitative study has been made of the effect of adsorbed gases on the interaction forces.

The present work has sought to provide a unifying thermodynamic approach to all these problems. Much of the earlier work on colloid stability was also based on a thermodynamic approach and some of the equations of the present paper have previously been applied to specific problems, though they were often derived by more intuitive methods which lack both the rigour and generality of a formal approach. There has, as far as we know, been no previous attempt to bring all the above problems together in a single general thermodynamic framework.

It is important to stress that colloidal phenomena are controlled both by thermodynamic and kinetic factors and that since different possible processes (e.g., the approach of two particles via Brownian motion, and the establishment of adsorption equilibrium at the particle surfaces) may occur on widely different time scales, the observed phenomena may correspond, in different

circumstances, to their evolution along different paths.   We
limit consideration here to the case in which adsorption equilibrium is maintained during the approach, encounter, and subsequent
aggregation or separation of two particles.   For simplicity we
deal with the case of the interaction between parallel plates.
        Rather than develop the theory in its full generality at
once, it is more convenient to treat first the case of solid
particles in a vapour phase: this reveals the more important
principles which are shown later to have quite general applicability.   We then consider interactions in non-electrolyte systems,
and finally the electrolyte case for both ideally polarizable and
non-polarisable electrodes.   In the latter case, we discuss two
specific examples: extensions to other systems are readily
devised.

Gas Adsorption on Parallel Plates

        The effect of gas adsorption on the force between interacting
plates has been discussed in an earlier paper(1) and we summarise
the analysis here.   An adsorptive gas and two parallel plates
are enclosed in a piston-cylinder arrangement as shown in
Figure 1 and the plates are positioned at an equilibrium separation by an external force Af (positive if the plates repel one
another) which is proportional to the surface area of each plate
and which in an infinitesimal movement dh contributes work −Afdh
to the system.   The differential Gibbs free energy of the system,
written relative to a reference system having the same temperature
bulk pressure and volume as those in Figure 1 but containing no
adsorbing plates, is shown to be

$$d(G - G^{\ominus}) = -(S - S^{\ominus})dT + \mu dn^{\sigma} + 2\sigma dA - Afdh \quad , \qquad (1)$$

where $n^{\sigma} = n - n^{\ominus}$ is the Gibbs adsorption of the vapour at a
chemical potential of μ and σ is the differential surface excess
free energy ('interfacial tension').   Integration of Equation (1)
at constant T, μ, σ and h with subsequent differentiation and
subtraction from (1) leads to a modified Gibbs adsorption equation

$$-2d\sigma = fdh + 2\Gamma d\mu \quad , \quad \text{(constant T)} \qquad (2)$$

where $\Gamma = n^{\sigma}/2A$ is the adsorbate surface excess concentration per
unit area of solid plate.   If Equation (2) is rewritten in the
limit of zero pressure, indicated by a superscript ^{o}, and subtracted from (2), we obtain

$$\Delta f = f - f^{o} = -2\left[\frac{\partial(\sigma - \sigma^{o})}{\partial h}\right]_{T,\mu} \quad , \qquad (3)$$

where Δf is the excess force over that between the plates in
vacuum.   Thus a change in the solid/gas interfacial tension with

plate separation contributes to the total force. Another more
useful expression for Δf emerges when the Maxwell relation between
f and Γ from Equation (2) is integrated with respect to the
fugacity of the adsorbate p*,

$$\Delta f = f - f^o = 2RT \int_o^{p*} \left(\frac{\partial \Gamma}{\partial h}\right)_{T,p} d \ln p* \quad \text{(constant T,h)} \quad , \quad (4)$$

where the fugacity is defined relative to the unit-fugacity ideal-
gas standard state. This equation presents the alternative but
equivalent view that changes in gas adsorption (at constant gas
pressure) with plate separation influence the total force.

The potential energy between the plates per unit plate area
and relative to zero at infinite plate separation is defined as
the integral of the force with respect to h at constant T and p.
Application of this definition to Equation (3) gives

$$v_p - v_p^o = \int_h^{\infty} (f - f^o)dh = 2\left[(\sigma - \sigma^o)_h - (\sigma - \sigma^o)_\infty\right] ,$$

$$\text{(constant T,p)} \quad , \quad (5)$$

where v_p^o is the potential energy of plate interaction in the zero
gas pressure limit and therefore arises only from the intermolecular forces between the plates. This vacuum potential energy
may be written approximately as $v_p^o = -A_H/12\pi h^2$ where A_H is
Hamaker's constant for the particular solid material($\underline{2}$), thus
enabling the total potential energy curve to be calculated from
knowledge of A_H and the dependence of the interfacial tension on h.

A more convenient relation for v_p follows by combining
Equations (4) and (5) and changing the integration order:

$$v_p - v_p^o = 2RT \int_o^{p*} \left[\Gamma(\infty) - \Gamma(h)\right]d \ln p* \quad \text{(constant T,h)} . \quad (6)$$

Thus the pressure dependence of $\left[\Gamma(\infty) - \Gamma(h)\right]$ (or the effect of h
on the adsorption isotherm) determines the effect of the gas
adsorbate on the potential energy of interaction.

Two opposing effects arise when the plates approach. First
the intermolecular potential fields emanating from each plate
overlap, causing an increase in the gas adsorption, and from
Equations (4) and (6), if this increase occurs at all pressures
an attractive component to the total interaction ensues. Second,
the diminishing adsorption space decreases the gas adsorption and
if this decrease occurs at all pressures, Equations (4) and (6)
indicate a repulsive component to the total interaction. The
principle that an overlapping of force fields furnishes an
attractive increment to the total potential energy whereas a
diminishing space furnishes repulsive increment is not specific
to gas adsorption but, as will be seen later, can be expressed in
more general terms.

These effects may be illustrated quantitatively by considering the adsorption of an ideal gas in the low pressure or
Henry's law region ($\underline{1}$). In this model the gas is distributed in

the intermolecular force-field between the plates, presumed to be
additive in the individual plate potentials, according to a
Boltzmann concentration profile(3) and the plate separation
dependence of Γ is determined by integrating the excess concentration profile over the available adsorption space. Application of
Equation (6) then permits evaluation of $v_p - v_p^o$. The results of
these calculations, using both (10:4) (A) and (9:3) (B) adsorption
potentials, are reviewed in Figure 2 (compare Figure 5, reference
1) in which the total potential energy curve v_p is shown as a
function of h.   To estimate the vacuum dispersion-force potential
v_p^o (solid line C) a Hamaker constant of 10^{-19} J was chosen,
corresponding approximately to the experimental value for mica
(4,5).   The dashed curve represents schematically the short range
repulsion when the plates get very close together.   Curves A and
B in the 0.8-0.9 nm (8-9 Å) range indicate the effect of the
adsorptive gas; they correspond to a pressure of 20 Nm^{-2} ($\sim$ 0.15
Torr), to a collision diameter of 0.34 nm (3.4 Å) and to a single-
plate adsorption well depth of 8 kT (approximately 22 kJ mol^{-1} at
room temperature).   Although no stabilising barrier is present in
Figure 2, a deep secondary minimum appears even at modest
adsorption energies, suggesting that the presence of an adsorptive
low pressure gas might cause loose flocculation of solid aerosols.
        At higher pressure, where multilayer adsorption commences,
the simple ideal gas model is invalid and no quantitative theory
is presently available to predict the variation of gas adsorption
with plate separation.   Nevertheless, qualitative arguments
suggest that a repulsive force contribution might occur at all
separations in this pressure region(1).   Adsorbate molecules in
the layers furthest away from the solid surface are not strongly
influenced by the solid plate potential field and will interact
with the outermost molecules adsorbed on the approaching second
plate before the plate potential fields overlap significantly.
This reduction of available adsorption space will decrease at
all separations and hence lead to a repulsive contribution to the
total potential energy.   The possibility therefore arises that at
higher gas pressures a potential barrier may exist to prevent
adhesion of the solid plates.   At pressures between the Henry
and multilayer regions intermediate behaviour should be expected
(e.g., see Figure 9 of Ref. 1).

Non-electrolyte Mixture Adsorption on Parallel Plates

        The analysis of the preceding section is readily extended to
the case of a c-component non-electrolyte liquid mixture confined
between parallel plates(1).   We review the essential features by
writing Equation (2) for a multicomponent system and by substitut-
ing the isothermal and isobaric Gibbs-Duhem equation of the bulk
liquid leading to

$$-2d\sigma = fdh + 2 \sum_{i=2}^{i=c} \Gamma_{i,1} d\mu_i \quad \text{(constant T,p)} , \qquad (7)$$

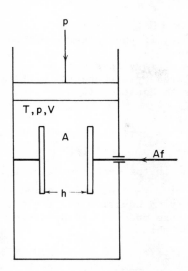

Figure 1.   Cylinder of volume V containing an amount n of gas at T,p; and two flat plates each of area A, separated by distance h, held in equilibrium by a force Af

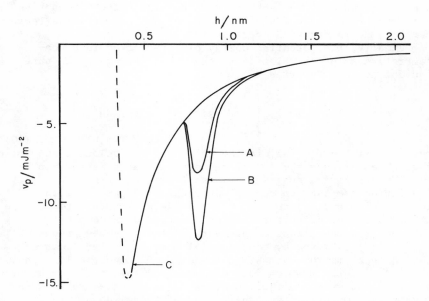

Figure 2.   Calculated potential energy curve for two plates in an adsorptive ideal gas A, for 10:4 potential; B for 9:3 potential at a pressure of 20 Nm⁻² (0.15 Torr) for single plate well depth of 8 kT, and collision diameter 0.34 nm (3.4 A); C van der Waals attraction (Hamaker constant 10⁻¹⁹ J); dotted line short range repulsion (schematic)

where $\Gamma_{i,1} = \Gamma_i - \dfrac{c_i}{c_1} \Gamma_1$ is the relative adsorption of species i

with respect to an arbitrary component 1.   Cross differentiation of Equation (7) followed by integration with respect to μ_i yields a relation analogous to Equation (4)

$$f_c - f_{c-1} = 2 \int_{-\infty}^{\mu_i} \left(\frac{\partial \Gamma_{i,1}}{\partial h}\right)_{T,p,\mu_{i \neq 1}} d\mu_i \quad (\text{constant } T,p,h,\mu_{j \neq i,1})$$

$$(8)$$

where f_c is the force in the c-component mixture and f_{c-1} is the force in the liquid mixture from which component i has been eliminated.   By applying Equation (8) (c-1) times to eliminate sequentially the mixture species and by adding the resulting c-1 force expressions, we obtain

$$f_c - f_1 = 2 \sum_{i=2}^{i=c} \int_{-\infty}^{\mu_i} \left(\frac{\partial \Gamma_{i,1}}{\partial h}\right)_{T,p,\mu_{i \neq 1}} d\mu_i \quad (\text{constant } T,p,h), (9)$$

where f_1 is the force between plates immersed in the pure liquid of species 1.   Equation (9) again reveals that the interaction force in a liquid mixture is influenced by the distance dependence of the adsorption isotherm and contains either attractive terms when $\Gamma_{i,1}$ increases with decrease in h or repulsive terms when $\Gamma_{i,1}$ decreases with decrease in h.

Finally, the potential energy of interaction in a non-electrolyte solution relative to the potential energy in pure liquid 1 is, cf. Equation (6),

$$v_p(c) - v_p(1) = 2 \sum_{i=2}^{i=c} \int_{-\infty}^{\mu_i} \left[\Gamma_{i,1}(\infty) - \Gamma_{i,1}(h)\right] d\mu_i$$

$$(\text{constant } T,p,h) \quad , \qquad (10)$$

where $\left[\Gamma_{i,1}(\infty) - \Gamma_{i,1}(h)\right]$ is the change in the relative adsorption of component i when the two plates approach at constant T,p and μ_i.   This relation suggests a means for calculating the effect of solution composition on the interaction of particles in a liquid medium and, therefore, may provide a path to estimate the variation of the effective Hamaker constant with composition. If we neglect the effect of pressure on liquid phase properties, $\left[v_p(1) - v_p^o\right]$ is given by Equation (6) when the upper limit of that integral is the saturation fugacity of component 1.   Thus addition of this limiting form of Equation (6) to (10) gives the potential energy in a liquid mixture $v_p(c)$ relative to that in vacuum v_p^o.   Quantitative application of these ideas must await further development in our understanding of dense gas and liquid mixture adsorption.

Electrolyte Adsorption on Ideally Polarized Plates

Early attempts to explain the effect of electrolytes on the
stability of lyophobic colloids culminated in the coherent picture
of the Derjaguin-Landau-Verwey-Overbeek (DLVO) theory(6,7).
Extension of the present formulation to this problem is straight-
forward(8) and provides rigorous expressions from which, on the
basis of a simple model, the DLVO theory is recovered.   Since
some of the earlier work, notably that of Overbeek(9), was also
based on a thermodynamic analysis, several of the expressions
derived below are closely similar to previously known equations.
The present work, however, reveals some implicit assumptions in
earlier discussions, provides a basis for further developments,
and completes the general account of the forces between colloidal
particles immersed either in electrolyte or non-electrolyte
solutions.   Since polarized and unpolarized particles behave
differently, we discuss each separately.

In the following, only an outline of theory is given: a
detailed account will be published separately(8).   By analogy
with the section on gas adsorption on parallel plates, two ideally
polarizable parallel plate electrodes are immersed in an electrolyte solution and enclosed in a piston-cylinder arrangement
(Figure 3).   The electrodes are held at a separation h by the
force Af, and are connected to an internal reference electrode,
with respect to which the plates can be held at a potential E.
For convenience we discuss the particular case of a system of
platinum electrodes, a dilute aqueous KCl electrolyte solution,
and, since it is reversible to the electrolyte anion, an internal
calomel reference electrode.   An external calomel reference
electrode containing KCl at a standard concentration $c^{\ominus}$ is also
included.   Potentials measured relative to this are denoted by
$E^{\ominus}$.   To describe the system, platinum ions and electrons are
chosen arbitrarily as the components of the electrode phases, and
potassium ions, chloride ions and water molecules as the components of the aqueous phase.   Equation (7) may be written in
terms of these components and combined (a) with the bulk electrode
and solution Gibbs-Duhem equations, (b) with the interfacial
electro-neutrality condition, $\sum_i z_i \Gamma_i = 0$ and (c) with the dissocia-
tion equilibrium conditions for KCl and Pt(10,11).   This then
yields the following form of the Gibbs adsorption isotherm:

$$-2d\sigma = fdh + 2qdE + 2\Gamma_{K^+,H_2O}\, d\mu_{KCl} \quad \text{(constant T,p)} \quad , \quad (11)$$

where Γ_{K^+,H_2O} is the relative adsorption of potassium ions with
respect to water at the platinum/solution interface, q is the
interfacial charge per unit area defined by

$$q = F(\Gamma_{Pt^+} - \Gamma_{e^-}) = F(\Gamma_{Cl^-} - \Gamma_{K^+}) \quad , \quad (12)$$

and E is the cell potential given by

$$FE = \bar{\mu}_{Cl^-} - \bar{\mu}_{e^-} + \mu_{Hg} - \tfrac{1}{2}\mu_{Hg_2Cl_2} \quad . \tag{13}$$

In Equation (13) F is the Faraday constant and $\bar{\mu}_i$, which depends on the electrical state of the phase in which i resides(12), is the electrochemical potential of charged species i; for neutral components $\bar{\mu}_i = \mu_i$, the ordinary chemical potential. When the two plates are infinitely separated, Equation (11) reduces to the classical Lippmann equation for a polarizable interface(11).

By arguments similar to those used previously, Equation (11) leads to several alternative expressions for the interaction force. If Equation (11) is written at the point (or potential) of zero charge (pzc) and, subtracted from Equation (11), we obtain

$$f(E) - f(E_{pzc}) = -2 \left[\frac{\partial(\sigma - \sigma_{pzc})}{\partial h} \right]_{T,p,E,\mu_{KCl}} , \tag{14}$$

which is analogous to Equation (3) except that here f at a cell potential E is relative to that in a solution of the same composition but at the potential of zero charge. Since the electrolyte concentration is constant $E - E_{pzc} = E^{\ominus} - E^{\ominus}_{pzc}$. Three other force expressions are available from the Maxwell relations of Equation (11)*. The cross-differential between f and q upon integration over the cell potential yields

$$f(E) - f(E_{pzc}) = 2 \int_{E_{pzc}}^{E} \left(\frac{\partial q}{\partial h} \right)_{T,p,\mu_{KCl},E} dE$$

$$(\text{constant } T,p,\mu_{KCl},h) \quad . \tag{15}$$

Variation of surface charge density with plate separation at constant cell potential and solution composition will affect the force.

Furthermore, the cross-differential between f and Γ, on integration with respect to the chemical potential of potassium chloride, gives the force in a solution having a chemical potential μ_{KCl}, relative to that in a standard reference solution at the same internal cell potential. When the concentration of potassium chloride is changed at constant E, the potential of the internal electrode, and hence that of the plates, changes relative to the external electrode. An increase in the concentration of potassium chloride c decreases the potential of the plates and by a suitable choice of c the plates can be brought to the potential of zero charge (pzc) which, relative to the external electrode is $E^{\ominus}_{pzc}$. We choose this concentration c_{pzc} as reference concentration, whence

---

* There are eight possible Maxwell equations involving f from
  Equation (11). We cite only three here.

$$f(\mu_{KCl}) - f(\mu_{KCl}^{pzc}) = 2 \int_{\mu_{KCl}^{pzc}}^{\mu_{KCl}} \left[\frac{\partial \Gamma_{K^+,H_2O}}{\partial h} \right]_{T,p,E,\mu_{KCl}} d\mu_{KCl}$$

$$\text{(constant T,p,E,h)} \quad , \tag{16}$$

where μ_{KCl}^{pzc} is less than μ_{KCl} for negatively charged plates and is greater for positively charged plates.

A third Maxwell relation results indirectly from Equation (11) by using the identity

$$\left(\frac{\partial E}{\partial q}\right)\left(\frac{\partial q}{\partial h}\right)\left(\frac{\partial h}{\partial E}\right) = -1 \quad ,$$

followed by substitution in the differential form of Equation (15) and integration with respect to surface charge:

$$f(q) - f(0) = -2 \int_0^q \left(\frac{\partial E}{\partial h}\right)_{T,p,\mu_{KCl},q} dq$$

$$\text{(constant T,p,}\mu_{KCl}\text{,h)} \quad , \tag{17}$$

where the lower limit of integration is again taken as the pzc.

These force expressions may be integrated according to Equation (5) to obtain the potential energy of interaction relative to that when the plates are at the pzc.

Thus Equation (14) yields

$$v_p(E,\mu_{KCl}) - v_p(E_{pzc},\mu_{KCl}) = 2\left[(\sigma - \sigma_{pzc})_h - (\sigma - \sigma_{pzc})_\infty\right]$$

$$\text{(constant T,p,h)} \quad . \tag{18}$$

Again the variation of the interfacial tension difference $(\sigma - \sigma_{pzc})$ with plate separation determines the potential energy of interaction (cf. Equation (5)). The difference $(\sigma_h - \sigma_\infty)$ is conventionally called the 'free energy' of the charged interacting plates (7,9).

Equation (15) with inversion of the order of integration gives

$$v_p(E,\mu_{KCl}) - v_p(E_{pzc},\mu_{KCl}) = 2 \int_{E_{pzc}}^E \left[q(\infty) - q(h)\right] dE$$

$$\text{(constant T,p,h)} \quad , \tag{19}$$

where it is assumed that E_{pzc} is independent of h. An equation similar to Equation (19) has been derived earlier both by an intuitive argument and by a less explicit thermodynamic treatment (7).

From Equation (16) we obtain for negatively charged plates

$$v_p(E,\mu_{KCl}) - v_p(E,\mu_{KCl}^{pzc}) = 2\int_{\mu_{KCl}^{pzc}}^{\mu_{KCl}} \left[\Gamma_{K^+,H_2O}(h) - \Gamma_{K^+,H_2O}(\infty)\right]d\mu_{KCl}$$

$$\text{(constant T,p,h)} \quad . \qquad (20)$$

The general principle enunciated in the earlier section on
gas adsorption on parallel plates again applies to the approach of
charged plates at constant internal cell potential and constant
bulk concentration of potassium chloride, provided attention is
focussed on the preferentially adsorbed ion (counter-ion).   Thus
for negatively charged plates Γ_{K^+,H_2O} is positive.   As the plates

approach the diffuse parts of the double layers begin to overlap
and since E is kept constant, positive ions are displaced from
between the plates; $\Gamma_{K^+,H_2O}(\infty) > \Gamma_{K^+,H_2O}(h)$ and the plates

experience a repulsive interaction.   Conversely, for positively
charged plates Γ_{K^+,H_2O} is negative and increases with decrease in

h, while Γ_{Cl^-,H_2O} is positive and decreases as the plates approach.

Thus, if the adsorption of the counter-ion decreases as the plates
approach, adsorption effects will make a repulsive contribution to
the potential of interaction.   Superimposed upon the purely
electrostatic interactions will be the intermolecular interactions
of the ions and the solvent which are usually neglected as a
second-order effect.

Finally, integration of Equation (17) gives

$$v_p(q,\mu_{KCl}) - v_p(0,\mu_{KCl}) = 2\int_0^q \left[(E(h) - E(\infty)\right] dq$$

$$\text{(constant T,p,h)} \quad , \qquad (21)$$

where $v_p(0)$ is again the potential energy at the pzc.   For ideally
polarizable plates, because no charge can be transferred across
the interface, it is possible to bring the plates together at constant charge, and at the same time to maintain an equilibrium ionic
distribution.   The distance dependence of the changing cell
potential at constant charge now determines the potential energy
of interaction.

A quantitative illustration of the thermodynamic formulation
is provided by the simple model of point charges (K^+ and Cl^- for
convenience) distributed in a uniform liquid dielectric confined
between two conducting parallel plates.   Since the point charges
interact only with the fields emanating from the charged plates,
the electrochemical potential of an ionic species may be rigorously
separated into a purely chemical and a purely electrostatic
part([12]),

$$\bar{\mu}_i = \mu_i + z_i F\phi \;,\tag{22}$$

where z_i is the ion charge and ϕ is the local electrostatic potential. This separation leads to a Boltzmann concentration distribution and upon combination with the Poisson equation from classical electrostatics provides a differential equation to describe the spatial variation of the local electrostatic potential between the interacting plates (13). For small electrostatic potentials the Poisson-Boltzmann equation is readily integrated and yields two expressions for the relation between the surface charge density q and the surface electrostatic potential ϕ_o depending on the boundary condition imposed at the plate surface. First, if the surface potential is held constant then the reduced surface charge density becomes

$$q_r(h_r) = \frac{q(h)}{\lambda Fc} = 2\xi_o \tanh\left(\frac{h_r}{2}\right) \quad (\text{constant } \phi_o)\;,\tag{23}$$

where c is the bulk molar concentration of potassium chloride, λ is the Debye length defined for a 1:1 electrolyte in a liquid solvent of uniform dielectric constant, $\xi_o = F\phi_o/RT$ is a reduced surface potential and $h_r = h/\lambda$ is the reduced distance between the charged plates. Secondly, if the surface charge density remains constant then the surface potential must vary according to

$$\xi_o(h_r) = \frac{q_r}{2} \coth(h_r) \quad (\text{constant } q)\;.\tag{24}$$

Finally, the reduced surface excess concentration (Γ_r) of potassium ions is

$$\Gamma_r = \frac{\Gamma_{K^+,H_2O}}{\lambda c} = -\xi_o \tanh\left(\frac{h_r}{2}\right)\;.\tag{25}$$

The reduced surface excess concentration of chloride ions is equal and opposite. Equations (22) to (25) supply the required results from the point-charge model.

Substitution of Equations (23) and (22), written once for chloride ions and once for electrons, and Equation (12) into Equation (19) yields

$$\frac{v_p(E,\mu_{KCl}) - v_p(E_{pzc},\mu_{KCl})}{2\lambda cRT} = \int_0^{\xi_o} [q_r(\infty) - q_r(h_r)]\,d\xi_o$$

$$= \xi_o^2\left[1 - \tanh\left(\frac{h_r}{2}\right)\right]$$

$$(\text{constant } T,p,h)\;.\tag{26}$$

This is closely similar to the DLVO result for small surface potentials, and becomes identical with it if it is assumed that $v_p(E_{pzc})$ arises solely from dispersion forces and contains no purely electrostatic contributions. It must also be stressed that the Hamaker constant to be used in calculating $v_p(E_{pzc})$ must

be that appropriate to plates separated by an electrolyte solution
of concentration c_{pzc} at the pzc.   A further limitation on the
classical result arises from the assumption involved in deriving
Equation (19), namely that E_{pzc} is independent of h.   While this
is true for the simple model, it will not hold if specific
adsorption of ions occurs, nor if the surface contains ionisable
groups.

For the restricted assumptions of the point-charge model at
small values of ξ_o, Equation (20) gives the same result as (26):
$v_p(E_{pzc}, \mu_{KCl})$ and $v_p(E, \mu_{KCl}^{pzc})$ are equal, since the Pt electrodes
both have the same potential $E_{pzc}^{\ominus}$, relative to the external
standard reference electrode when the pzc is independent of c.

As shown in Figure 4 the plates experience a repulsion at all
distances.   It may be noted that whereas in the case of gas
adsorption, repulsion forces result from molecular size effects,
in the case of electrolytes the repulsion arises, in the simplest
case, from electrostatic forces: ion size effects may also appear
as a second contribution.

Application of the point-charge model to Equation (21) provides a constant-charge potential energy of interaction which,
because of the different integration path, is distinct from
Equation (26).   From Equations (13), (21), (22) and (23) we
obtain

$$\frac{v_p(q, \mu_{KCl}) - v_p(0, \mu_{KCl})}{2 \, cRT} = \int_0^{q_r} \left[\xi_o(h_r) - \xi_o(\infty)\right] dq$$

$$= \xi_o^2(\infty) \left[\coth\left(\frac{h_r}{2}\right) - 1\right]$$

$$\text{(constant } T, p, h) \quad . \tag{27}$$

This result is that of the DLVO theory applied to colloidal
particles which collide with a constant surface charge(14).

To maintain constant charge when two charged polarizable
plates approach, the effect of overlapping of the diffuse parts of
the double layers which tends to decrease the adsorption of the
counter-ions has to be off-set by an increase in the potential.
Thus, from Equation (27), a repulsive potential energy is
generated.   We note (Figure 4) that the repulsive effect is more
pronounced for the constant charge integration path and that at
small separations the constant-charge potential energy approaches
infinity.   Since the present linearised point-charge model
equations apply only for small values of the surface potential,
Equation (27) must break down for separation distances near zero.

Electrolyte Adsorption on Ideally Non-polarizable Plates

To discuss the interaction between reversible (i.e., ideally
non-polarizable) particles the platinum electrodes in Figure 3 are
replaced by silver/silver chloride electrodes.   Since silver and

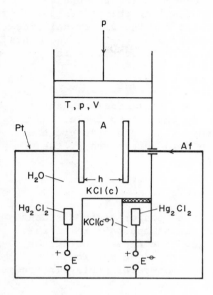

Figure 3.   *Cylinder containing a symmetric aqueous electrolyte (potassium chloride) and two plates charged to a potential E relative to an internal reference calomel electrode and to a potential E⊖ relative to an external standard calomel electrode*

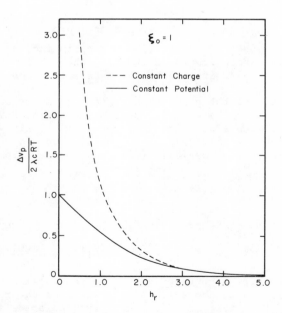

Figure 4.   *Calculated effect of a dilute point-ion electrolyte on the potential energy for two charged plates from Equations 26 and 27*

chloride ions are present in the aqueous solution to an extent
determined by the solubility product of silver chloride K_{AgCl} they
both serve as potential determining ions for the silver chloride
electrodes. Following the earlier procedure, the modified Gibbs
adsorption equation for the reversible electrode system is

$$-2d\sigma \;=\; fdh \;+\; 2\Gamma_{K^+,H_2O}\, d\mu_{KCl} \quad (\text{constant } T,p) \;. \tag{28}$$

Comparison of Equation (28) with Equation (11) discloses the
absence of the cell potential as an independent variable. For
reversible electrodes at constant temperature and pressure the
cell potential cannot be varied since E is determined by the
chemical potentials of pure phases only:

$$FE \;=\; \mu_{Hg} \;+\; \mu_{AgCl} \;-\; \mu_{Ag} \;-\; \tfrac{1}{2}\mu_{Hg_2Cl_2} \;. \tag{29}$$

Because the internal cell potential is no longer an independent
variable, non-polarizable electrodes cannot approach reversibly
under the condition of constant charge. By repetition of the
thermodynamic development given in the preceding section, the
potential energy of interaction between non-polarizable electrodes
becomes

$$v_p(\mu_{KCl}) - v_p(\mu_{KCl}^{pzc}) \;=\; 2\int_{\mu_{KCl}}^{\mu_{KCl}^{pzc}} \left[\Gamma_{K^+,H_2O}(h) - \Gamma_{K^+,H_2O}(\infty)\right] d\mu_{KCl}$$

$$(\text{constant } T,p,h) \quad, \tag{30}$$

which is analogous to Equation (20) and again indicates the role
of electrolyte adsorption in determining the electrostatic part of
the total potential energy curve for interacting reversible
electrodes.

    Application of the point-charge model to Equation (30) yields
a result similar to that of the constant potential DLVO theory.
Substitution of the electrochemical potential equality condition
of the chloride (or silver) ions in the aqueous and solid silver
chloride phase, Equations (22), (25), and the relation that

$$d\mu_{KCl} \;=\; RT\, d\ln\left[\frac{c}{2}\left(1 + \sqrt{1 + \frac{4K_{AgCl}}{c^2}}\right)\right]$$

gives, as a first order approximation of a series expansion,

$$\frac{v_p(\mu_{KCl}) - v_p(\mu_{KCl}^{pzc})}{2\lambda cRT} \;=\; \xi_o \left[1 - \tanh\left(\frac{h}{2}r\right)\right]\left[1 + \frac{K_{AgCl}}{c^2}\right] \;. \tag{31}$$

We notice, however, a correction term to the constant potential
DLVO theory involving the solubility product of the reversible
electrodes. For substances such as silver chloride (solubility
product $\sim 10^{-10}(\text{mol dm}^{-3})^2$ the correction term is unimportant for
added KCl concentrations of $> 10^{-4}$ mol dm^{-3}. The correction

factor will not be negligible for more soluble substances, e.g.,
$MgCO_3$, $K_s = 2.6 \times 10^{-5}$ when c is less than 5×10^{-2} mol dm^{-3}.
The present thermodynamic treatment thus provides a means for
extending the DLVO theory to include interactions between more
soluble dispersed solids.

Conclusions

A general thermodynamic treatment is presented to describe
the effect of adsorption of gases and mixtures of both electrolytes and non-electrolytes on the interaction of solid particles.
The formulation provides a coherent picture of the effect of
adsorption in terms of the opposing tendencies of the overlap of
the interparticle force fields (intermolecular or electrostatic)
and of the availability of adsorption space.   The two idealised
cases of polarizable and non-polarizable electrodes are employed
to describe the interaction of charged particles in an electrolyte
medium.   A quantitative point-charge model applied to both types
of interacting particles yields, under different conditions, the
constant potential and constant charge Derjaguin-Landau-Verwey-
Overbeek theories.   Extension of the DLVO theory to the case of
less sparingly soluble solids is outlined.

Literature Cited

1. Ash, S.G., Everett, D.H. and Radke, C.J., J.Chem.Soc.Faraday
        Trans.II, (1973), 69, 1256
2. Hamaker, H.C., Physica (1937), 4, 1058
3. Barker, J.A. and Everett, D.H., Trans.Faraday Soc., (1962),
        58, 1608
4. Tabor, D. and Winterton, R.H.S., Proc.Roy.Soc. (1969), 312A,
        435
5. Israelachvili, J.N. and Tabor, D., Proc.Roy.Soc., (1972),
        331A, 19
6. Derjaguin, B.V. and Landau, L., Acta Physicochem.USSR (1941)
        14, 633
7. Verwey, E.J. and Overbeek, J.Th.G., "Theory of the Stability
        of Lyophobic Colloids", Elsevier, Amsterdam, 1948
8. Everett, D.H. and Radke, C.J., in preparation
9. Overbeek, J.Th.G., in "Colloid Science", H.R.Kruyt, Ed.,
        Vol. I, Chap. 4,6, Elsevier, Amsterdam, 1952
10. Parsons, R., "Thermodynamics of Electrified Interfaces", in
        'Source Book of Colloid and Surface Chemistry',
        H. van Olphen, Ed., to be published
11. Newman, J., "Electrochemical Systems", Prentice-Hall,
        Englewood Cliffs, New Jersey, 1973
12. Guggenheim, E.A., "Thermodynamics", 5th ed., North-Holland,
        Amsterdam, 1967
13. Parsons, R., in "Modern Aspects of Electrochemistry",
        J.O'M. Bockris, Ed., Vol. 1, Chap. 3, Academic Press,
        New York, 1954
14. Usui, S., J. Colloid Interface Sci., (1973), 44, 107

2

Configurational Behavior of Isolated Chain Molecules Adsorbed from Athermal Solutions

M. LAL and M. A. TURPIN
Unilever Research, Port Sunlight, Cheshire L62 4XN, England

K. A. RICHARDSON
Department of Polymer and Fibre Science, University of Manchester, Institute of Science and Technology, Manchester, England

D. SPENCER
Unilever Computing Services Ltd., Bromborough Port, Cheshire, England

Models which have been considered in the various analytical treatments of statistical thermodynamic and configurational behaviour of adsorbed polymer layers most often ignore such important characteristics of the adsorbate molecules as the excluded-volume effect and bond rotational hindrance (1, 2, 3, 4, 5). Further, the procedures used for introducing solvent and concentration effects are less satisfactory (6). The present analytical developments are thus limited to oversimplified models that would have only a little correspondence with real systems. It is, therefore, desirable to investigate alternative approaches which would be effective in the study of more realistic models.

The success of Monte Carlo computer simulation method in investigating the configurational and thermodynamic behaviour of systems involving polymer molecules assuming models of varying complexity is well established (7, 8, 9). This approach is based on the assumption that it is possible to generate a random sample of molecular configurations which would adequately simulate the canonical ensemble corresponding to the model assumed. Then the mean values of various properties over such a sample would converge to the canonical ensemble averages. The extension of the Monte Carlo method to polymers interacting with interfaces has so far been carried out only to a limited extent (10, 11). This is because of a serious difficulty pertaining to sample convergence that one would encounter in the case of strongly interacting systems, if the usual techniques are used. The first objective underlying the present work is to introduce a more sophisticated Monte Carlo scheme in this area with the aim of overcoming the difficulty just mentioned. The successful application of such a scheme should lead to reliable studies on the models which adequately take into account the solvent and concentration effects. This is the first of a series of papers on our

studies of the configurational state of chain molecules in the
adsorbed state. Here we describe the present approach and
present its application to an isolated excluded-volume chain
interacting with a surface. This model may be considered to
represent a polymer molecule adsorbed from an extremely dilute
athermal solution.

The Present Approach

The equilibrium value of a quantity, q, $\bar{q}$, can be
identified with the canonical ensemble average:

$$\bar{q} = \frac{\int q \, \exp(-U/KT) \, d\Omega}{\int \exp(-U/KT) \, d\Omega}$$

where $\exp(-U/KT)$ is the Boltzmann factor corresponding to a
value q of the quantity. The integrations in the above
equation extend over the total configurational phase space,
Ω, of the system.

Let us draw a random sample of configurations from the
phase space Ω. The mean of the quantity q, <q>, over such
a sample is given as

$$<q> = \frac{\sum^{S} q_i \, \exp(-U_i/KT)}{\sum^{S} \exp(-U_i/KT)}$$

where S is the magnitude of the sample. The Monte Carlo
approach assumes that for large S, <q> would converge to $\bar{q}$.
For systems involving strong interactions, sampling procedures
based on purely random collection of configurations will not
be satisfactory as to achieving convergence that will produce
estimates within reasonable limits of uncertainty. The way to
obtain samples with acceptable convergence in such cases is to
utilise a sampling scheme which favours the high density
regions of the phase space. In a previous publication we
presented a scheme suitable for tackling chain molecular
models involving inter-segmental interactions (12). This
scheme can be modified so as to be congenial to the model
under the present consideration.

The present method involves generating a homogeneous
markov sequence of configurations with one-step transition
probabilities (p_{ij}) given such that the configuration sample
thus obtained assumes Boltzmann distribution in the limit of
a long sequence. Hence the simple means over such a sample
can be identified with the canonical ensemble averages. The
desired configuration sequences can be generated following the
commonly used asymmetric procedure according to which the
transition probabilities are given as (13)

$$p_{ij} = 1 \text{ for } U_j \leqslant U_i$$

$$p_{ij} = \exp(-U_j/KT)/\exp(-U_i/KT) \text{ for } U_j > U_i$$

$$p_{ii} = 1-p_{ij}$$

The sequence can start from any allowed configuration chosen arbitrarily. Let the energy corresponding to this configuration be U. Consider another configuration, with energy $'U$, which can be derived from the previous one by some random process. If $'U$ is less than or equal to U, accept the new configuration in the sequence. Else, choose a random number, x, between O and 1 and compare it with $r = \left[\exp('U/KT)/\exp(-U/KT) \right]$. If $r \geqslant x$, also then the new configuration assumes the next position in the sequence. Otherwise, reject the new configuration letting the system return to the preceding one, which means that the same configuration is counted again in the sequence. Repeat the above process to acquire further moves. In this way the sequence can be extended to any length. In the previous scheme the mechanism by which the system moved from one configuration to another involved the rotation of a single bond in the chain. This resulted in the changes in the positions of all the segments succeeding the rotating bond. This scheme, although successful in case of free chains, is unsuitable in the presence of an interface. For the present model we developed a mechanism based on the consideration that transition from one configuration to another occurs as a consequence of changes in the rotational positions of only a few consecutive bonds. Let us term such a sequence of bonds as the transition region (figure 1). The position of the transition region in the molecule is assumed to be stochastic in nature. In the present scheme the number of bonds constituting the transition region is four. This is the minimum size of the region which must be considered in order to generate all the possible configurations which can be assumed by a linear chain molecule confined to a tetrahedral lattice. We further assume that beyond the transition region no changes in the positions of the segments will occur.

In order that the samples achieve the required convergence, it is necessary that the ergodicity and steady state conditions are met. In simple terms, the ergodicity condition stipulates that starting from a configuration it is possible to reach any permissible configuration in a finite number of steps, and that the sample contains no periodicities. The steady state condition states that

$$\sum_i u_i \, p_{ij} = u_j \quad \text{for all } j$$

where u_i and u_j are the absolute probabilities of the occurrence of states i and j in the sample. We reckon that the samples generated by our scheme are in reasonable accord with the above criteria.

The Model

The geometry corresponding to the four valence bonds of the carbon atom implies that polymer molecules with -C-C-C- "back-bone" can be adequately represented by the tetrahedral chains. Our model is thus based on a tetrahedral lattice in which the lattice sites are occupied either by chain segments or by solvent molecules. The value of the C-C $\diagup$C bond angles in such chains is $109.28°$, and the values of the rotational angles which the bonds can assume are $0°$ (trans), $120°$ (gauche$^+$) and $240°$ (gauche$^-$). The lattice model assumes that the above values of the bond angles and the bond rotational angles are invariant. This approximation seems not unreasonable. It is necessary to stipulate that the chains will assume only those configurations which do not contain multiple occupancies (the excluded-volume condition). The adsorbent is represented by a two-dimensional surface in the lattice with the adsorption sites located at the lattice points.

In solutions there exist three kinds of interactions: (1) segment/segment interactions, (2) solvent/solvent interactions and (3) segment/solvent interactions. Let ϵ_{11}, ϵ_{22}, ϵ_{12} denote, respectively, the energies associated with the segment/segment, solvent/solvent and segment/solvent pairs. The energy change accompanying the formation of a segment/segment pair, $\Delta\epsilon_1$, will be

$$\Delta\epsilon_1 = \epsilon_{11} + \epsilon_{22} - 2\epsilon_{12}$$

$\Delta\epsilon_1$ serves as the solvent parameter in the model. Good solvents oppose the formation of segment/segment pairs, which implies that for such solvents $\Delta\epsilon_1$ assumes positive values. In bad solvents the formation of segment/segment pairs will be favoured, hence $\Delta\epsilon_1$ will be negative. $\Delta\epsilon_1 = 0$ corresponds to athermal solutions.

The second energy parameter characterising our model is the segment/surface binding energy $\Delta\epsilon_2$, which is defined as

$$\Delta\epsilon_2 = \epsilon_{1S} - \epsilon_{2S}$$

where ϵ_{1S} is the energy associated with a segment/surface bond and ϵ_{2S} is that associated with a solvent/surface bond. Thus $\Delta\epsilon_2$ is the energy change due to the formation of a segment/surface bond at the expense of a solvent/surface bond. The adsorption can only take place if $\Delta\epsilon_2$ is negative.

Computation

A computer program was developed for generating configuration sequences following our scheme assuming the model described in section 3. The program proceeded as follows:

1. Read the following data:

    N = total of segments in the chain
    M = length of the configuration sequence to be generated
    L = length of interval in the sequence corresponding to
        the output of data in intermediate stages
    $\Delta\epsilon_1, \Delta\epsilon_2$: the energy parameters.

2. Generate an initial configuration $[x_1, y_1, z_1; x_2, y_2, z_2; \ldots$
    $\ldots; x_i, y_i, z_i; \ldots; x_N, y_N, z_N]$; (n_i, y_i, z_i)
    define the position of a segment i.

3. Define the adsorbent plane such that y = x+ constant.

4. Determine the various quantities of interest (e.g. number of segments on the surface, square of the end-to-end distance, square of the radius of gyration, etc.).

5. Generate a new configuration by randomly selecting four consecutive bonds in any part of the chain and moving these bonds to an alternative configuration, whilst the remainder of the molecule is unaltered.

6. Check the new configuration for excluded-volume. If the configuration involves multiple occupancy of any lattice site, go to 14.

7. Calculate ΔU, the difference between the energies of the previous and the new configuration.

8. If ΔU is positive or zero, go to 13.

9. Calculate r = exp(ΔU/KT).

10. Generate a random number (x) between 0 and 1.

11. If r $\geqslant$ x, go to 13.

12. Go to 14.

13. Calculate the desired quantities for the new configuration.

14. Accumulate the calculated quantities.

15. If the number of configurations generated $\neq$ AL (where A is any integer), go to 5.

16. Print the desired information.

17. If the number of configurations generated <M, go to 5.

18. Calculate the simple averages of the various quantities over the sample.

19. Print the computed averages and other relevant information.

The calculations were carried out on a CDC 7600 computer. Convergence was assumed to have been reached if the cumulative averages of the various quantities did not vary beyond certain set limits as the sample size increased. The sample sizes required for reasonable convergence exceeded 500,000 configurations. It took approximately ten minutes to generate a sample of such a magnitude on the above machine. If a sample did not show satisfactory convergence, the calculation was restarted from the last configuration generated and its size gradually increased till the required convergence was attained.

Results and Discussion

An outstanding feature of the configurational state of a polymer molecule interacting with a surface is that, except for very high $-\Delta \epsilon_2$, not all the segments are found to be in the adsorbed state. This state is pictorially represented in figure 2, where it is shown that the configuration is constituted of "trains" (sequences of consecutive segments in the adsorbed state), "loops" (sequences of consecutive segments in the unadsorbed state) and "tails" (unadsorbed arrays of segments at the ends). Basic quantities relating to the above configurational state are the fraction of segments in the adsorbed state (ν) and thickness of adsorbed layer (τ). Additional parameters such as those pertaining to the "train", "loop" and "tail" sizes and the height distribution of segments above the surface would provide a more detailed knowledge of the configurational state.

The present calculations have been directed towards the evaluation of the various configurational parameters noted above. In this study we have considered a model constituted of a linear chain molecule of 100- segment length adsorbed from a solvent for which $\Delta \epsilon_1 = 0$, which implies that for the present system the solvent/solute interactions are invariant with respect to the intramolecular configurational changes. Such solutions are known as athermal solutions.

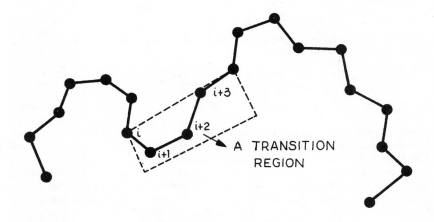

Figure 1.   A transition region is constituted of four consecutive bonds (i, $i + 1$, $i + 2$, $i + 3$) in any part of the chain

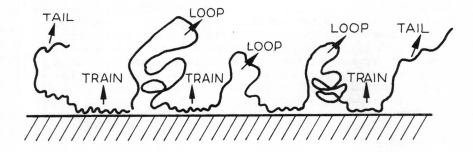

Figure 2.   The train–loop–tail model for an adsorbed polymer molecule

The mean fractions of segments bound to the surface calculated assuming various values of segment/surface binding energy are presented in figure 3. We observe that $<\nu>$ increases rapidly with the increase in $(-\Delta\epsilon_2/KT)$: at $-\Delta\epsilon_2/KT = 0.5$, less than 10% of the segments are adsorbed, while at $-\Delta\epsilon_2/KT = 1.4$, approximately 75% of the segments would be on the surface. The $<\nu>$ versus $-\Delta\epsilon_2/KT$ plot can be extrapolated to determine the point below which $<\nu>$ is zero. Such a point is known as the critical point, and the corresponding energy is the critical energy of adsorption. For the present system $(-\Delta\epsilon_2/KT)_c$ is found to be ~0.45. To our knowledge no calculations have heretofore been reported on a model identical to the one under the present consideration. However, McCrackin carried out Monte Carlo calculations on the systems involving terminally anchored chains constrained to a cubic lattice (11). The behaviour of $<\nu>$ with regard to its variation with the increase in the energy of adsorption found by McCrackin is qualitatively similar to that given by figure 3. There is, however, appreciable quantitative difference in the two results : the critical energy for the terminally anchored chains considered by McCrackin was found to be approximately 0.25 KT, which is considerably lower than that obtained for the present system; also, our values of $<\nu>$ tend to be lower than those which would be obtained by extrapolating McCrackin's results to the corresponding adsorption energies and the chain length. This may be due to the reason that a "train" in tetrahedral chain can exist only in a single configuration ---tttt---. Hence it is in the state of zero configurational entropy. This means that on adsorption a sequence of segments will have to lose whole configurational entropy associated with it in the solution phase. The adsorption is, thus, only possible if the segment/surface interaction energy is so large that it would more than compensate the entropy loss. Secondly, systems such as those studied by McCrackin have already lost some entropy on account of the additional constraint of one chain-end permanently anchoring to the surface. Thus the segment/surface energy required for adsorption to take place would be lower, as it has lower entropy to overcome.

Figure 4 gives the mean fractions of segments in the successive layers above the surface computed for various values of $-\Delta\epsilon_2$. We note that in the case of $\Delta\epsilon_2 = -0.5$ KT the distribution is markedly different from those obtained for higher segment/surface binding energies. For $\Delta\epsilon_2 = -0.5$ KT, the segment density varies little in the first few layers adjacent to the surface, and then declines slowly as the distance from the surface increases. For higher interaction energies, on the other hand, the initial decrease in the density with distance is very rapid, which subsequently slows down as the density approaches zero. The thickness of adsorbed

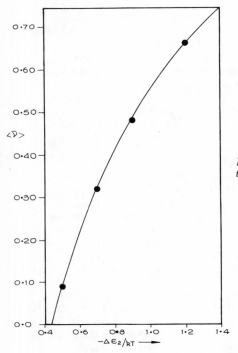

Figure 3.   *Mean fraction of segments bound to the surface ($<v>$) vs. energy of adsorption per segment ($-\Delta\epsilon_2/kT$)*

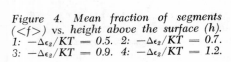

Figure 4.   *Mean fraction of segments ($<f>$) vs. height above the surface (h). 1: $-\Delta\epsilon_2/KT = 0.5$. 2: $-\Delta\epsilon_2/KT = 0.7$. 3: $-\Delta\epsilon_2/KT = 0.9$. 4: $-\Delta\epsilon_2/KT = 1.2$.*

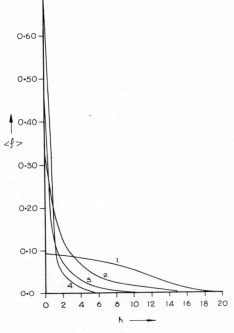

layer (τ) can be defined as the distance from the surface at which the segment-density vanishes, viz. the distance at which a distribution curve in figure 4 meets the abscissa. The values of $\langle \tau \rangle$ thus determined are plotted against $-\Delta \epsilon_2/KT$ in figure 5. Thickness of adsorbed layer is found to be a rapidly decreasing function of the segment/surface interaction energy. The magnitude of $\langle \tau \rangle$ at $\dfrac{-\Delta \epsilon_2}{KT} = 0.5$ ($\sim$20 units) suggests that when the interaction energy is small, the molecule in the adsorbed state would assume dimensions similar to those in solution. But at energies exceeding -1.0 KT the adsorbed molecule would be considerably flattened, as is indicated by the small values of $\langle \tau \rangle$.

Analysis of the size distributions of "trains", "loops" and "tails" reveals that the "train" and "loop" sizes assume narrow distributions whereas the "tail" size distributions are extremely wide showing large fluctuations. The mean sizes of "trains" ($\langle L_t \rangle$), "loops" ($\langle L_\ell \rangle$), and "tails" ($\langle L_e \rangle$) calculated at several values of $\Delta \epsilon_2$ are plotted in figure 6. The mean "train" size varies between three and six in the energy range considered. At $\Delta \epsilon_2 = -0.5$ KT, the mean "loop" size is as large as about 20 units but follows a sharp decrease as $-\Delta \epsilon_2$ increases. As expected, the mean "tail" length decreases with the increase in $-\Delta \epsilon_2$. The interesting feature which our analysis brings out is that, except at very high $-\Delta \epsilon_2$, a large proportion of unadsorbed segments constitute "tails", and that the "tails" contribute most significantly to the thickness of adsorbed layer.

Picture of an isolated adsorbed molecule emerging from the present study is that at low surface/segment interaction energies only a small fraction of the segments is bound to the surface, and the shape and dimensions of the adsorbed molecule are comparable to those assumed in solution. At increased values of the interaction energy, however, the number of segments on the surface is significant, and the configurational state assumed is such that small "trains" and small "loops" are favoured while "tail" lengths are relatively large. Most important contribution to the thickness of adsorbed layer derives from "tails".

Concluding Remarks

In this study we have deliberately assumed an extremely simple model, for our main purpose here was to ascertain the viability of the Metropolis sampling method in the case of chain molecular systems interacting with a surface. Notwithstanding the large magnitude of the samples required for satisfactory convergence, the Monte Carlo approach considered may be regarded as successful in tackling the present systems. We hope that this method proves equally successful when

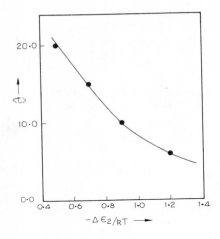

Figure 5. *Mean thickness of adsorbed layer ($<\tau>$) as a function of the energy of adsorption per segment*

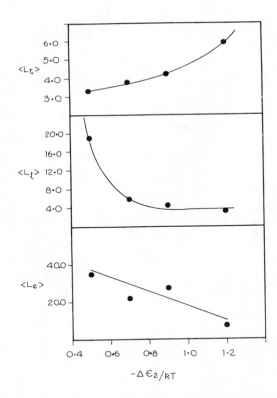

Figure 6. *Mean values of "train" ($<L_t>$), "loop" ($<L_l>$), and "tail" ($<L_e>$) lengths corresponding to various values of the adsorption energy per segment*

modifications such as the trans/gauche bond conformational energy difference and non-athermal solvents are introduced in the model. We shall deal with such models in further studies.

An important extension of this work that we envisage concerns the investigation of the configurational behaviour of adsorbed polymer molecules as the surface coverage varies. This can be achieved by introducing periodic boundaries in the system (14).

Literature Cited

1. Frisch, H.L., Simha, R., and Eirich, F.R., J. Chem. Phys., (1953), 21, 365.

2. Hoeve, C.A.J., DiMarzio, E.A., and Peyser, P., J. Chem. Phys., (1965), 42, 2558.

3. Rubin, R.J., J. Chem. Phys., (1965), 43, 2392.

4. Roe, R.J., Proc. Natl. Acad. Sci., (1965), 53, 50.

5. Motomura, K., and Matuura, R., J. Chem. Phys., (1969), 50, 1281.

6. Silberberg, A., J. Chem. Phys., (1968), 48, 2835.

7. Lal, M., R.I.C. Rev., (1971), 4, 97.

8. Lal, M., and Spencer, D., J. Chem. Soc. Faraday Trans. II, (1973), 69, 1502.

9. Lal, M., and Spencer, D., J. Chem. Soc. Faraday Trans. II, (1974), 70, 910.

10. Bluestone, S., and Cronan, J. Phys. Chem., (1966), 70, 306.

11. McCrackin, F.L., J. Chem. Phys., (1967), 47, 1980.

12. Lal, M., Mol. Phys., (1969), 17, 57.

13. Wood, W.W., in "Physics of Simple Liquids", pp 117-230, North Holland Publishing Co., Amsterdam, 1968.

14. Clark, A.T., and Lal, M., to be published.

3

Equilibrium Film Pressure on a Flat, Low-Energy Solid

ROBERT J. GOOD

Department of Chemical Engineering, State University of New York at Buffalo, Buffalo, N. Y. 14214

Introduction

The equilibrium film pressure of an adsorbate, π_e, is defined as

$$\pi_e = \gamma_S - \gamma_{SV} \tag{1}$$

where γ_S is the surface free energy of the solid in vacuum, and γ_{SV} is the surface free energy of the solid in contact with the saturated vapor of the adsorbate. The value of this property for an adsorbate which, as a liquid, forms a non-zero contact angle on the solid, has been a matter of uncertainty for some time ([1]-[9]). This fact has detracted from the usefulness of measurements of contact angle, θ, for the estimation of solid surface free energies ([2],[3]). See refs. ([5]) and ([6]) for an important discussion of the problem as it has existed in recent years.

The measurement of π_e is not particularly easy; and up to very recently ([8]) the only determinations that had been reported for low-energy solids were made on powders ([4],[5],[6]), while reported contact angle measurements were made on essentially flat surfaces.

Fox and Zisman ([1]) found reason to conclude that π_e is probably negligible; and this assumption is basic to their "γ_c" method of treating contact angle data. Adamson ([3]) has pointed out that the existence of a thick adsorbed film, and consequent non-negligible value of π_e, are compatible with the existence of a non-zero contact angle, provided the material in the adsorbed film has a structure that is significantly different from the bulk liquid. He hypothesizes a degree of "structure" in a multilayer film, which decays with distance from the surface. If such thick, structured multilayers exist, the low

28

entropy associated with such a structure could account
for the existence of a non-zero contact angle. But of
course, one cannot argue that the existence of a non-
zero contact angle proves the presence of a <u>thick</u>,
structured film. And indeed, it is hard to <u>imagine</u>
that a multilayer film of, say, CCl_4 on Teflon could
have a "structure" that would meet the requirements
noted in Ref. 3. Whalen (<u>6</u>) pointed out that the
Hill-deBoer equation (<u>11</u>,<u>12</u>,<u>13</u>), with empirical constants obtained from measurements with hexane and
octane on Teflon TFE in the low-coverage region, predicts a submonolayer limiting adsorption at saturation.
    The purpose of this paper is to develop a method
of estimating π_e, <u>a priori</u>, on a molecularly flat,
homogeneous, low-energy surface, for adsorbates of
liquids for which $\theta > 0$, without introducing any
physical interactions other than those for which
quantitative expressions are available in well-known
physical theory (<u>10</u>).
    As a first step in this analysis, an attempt was
made to compute, by methods which will be described
below, the constants for a BET multilayer adsorption
isotherm for CCl_4 on a homogeneous, molecularly
smooth fluorocarbon surface. It was found that the
BET "c" constant was very small. If the adsorbent
were a powder instead of a flat surface, a very low
value of c would indicate a BET type III isotherm,
the final up-turn of which (near $p = p_0$) was due to
enhanced adsorbate-solid interactions at points of
solid-solid contacts, e.g. as "pendular rings", which
Wade and Whalen (<u>5</u>,<u>6</u>) have discussed. Such structural
features are, by definition, absent from a molecularly
smooth surface.
    These preliminary results led us to turn to a
Langmuir model, as being far easier to treat than a
BET model. It was anticipated that a strictly Lang-
muirian model might break down when some possible
values of molecular parameters were assumed — for
example, the model might predict high enough coverage
that lateral interactions would be important. For
such a regime, a Hill-deBoer isotherm would be more
suitable. However, in the present computations, we
confined that aspect of the study to the estimation
of the conditions under which the Langmuir postulates
break down. To anticipate some of our results, a
breakdown of the Langmuir assumptions was in fact
found, and in a very interesting region of the range
of chain-length for n-alkanes, on Teflon.

Theory

Consider the low-coverage sub-monolayer region of an adsorption isotherm on a uniform solid. We will estimate the energy and entropy of adsorption, with the bulk liquid as the reference state. The partial molal entropy of the adsorbed molecules, in a Langmuir adsorbed film on a solid, is given by (14).

$$\overline{S} = - R\ln\left[x_a/(1-x_a)\right] \qquad (2)$$

where x_a is the "θ" of the Langmuir adsorption equation, i.e. the mole fraction of surface sites that are occupied. This entropy term is configurational; it arises from the possibility of permuting the molecules among the occupied and vacant sites. The partial molal entropy of transfer, from bulk liquid to adsorbed film, is given by,

$$\overline{\Delta S} = - R\ln\frac{x_a}{1-x_a} + R\ln\frac{W_a}{W_L} + \Delta S_{internal} - S_{ql} \qquad (3)$$

The last two terms in Equation 3 can probably be neglected. These correspond, respectively, to the changes in internal, molecular degrees of freedom, and to the configurational entropy of the liquid, considered as a quasi-lattice with occupied and vacant sites. The latter term is small, probably less than 0.5 entropy unit (15).

The second term on the right involves the number of angular configurations accessible to a molecule in the liquid (W_L) and in the adsorbed state (W_a). For a quasi-spherical molecule such as CCl_4, this term is zero. For an elongated molecule, such as an n-alkane longer than propane or n-butane, we may estimate this term as follows: The molecule is treated as a rigid cylinder, of length l and diameter d. In the liquid, its axis may be at any angle, relative to a fixed coordinate system; in effect, it has a volume $0.75\pi(l/2)^3$ accessible to it. In the adsorbed state, the energy of attraction between the solid and the extended molecule renders it improbable that the molecule will have its axis in any plane that is at an appreciable angle to the surface. So it may be regarded as having, accessible to it, a volume $2\pi(l/2)^2 d$. This entropy term, then, is approximately $R\ln(3d/4l)$. A further refinement on this term can be made by counting the numbers of bent configurations explicitly. These are tedious to enumerate, but their effect on the equilibrium will be to predict even lower coverage than that estimated with their neglect.

Thus, we can write as a reasonable approximation, for
the entropy of transfer per mole between liquid and
adsorbed state:

$$\overline{\Delta S} \cong - Nk[\ln\frac{x_a}{1-x_a} - \ln(\frac{3d}{4\ell})] \qquad (4)$$

where N is the Avogadro number. For symmetrical adsorbate molecules, the last term in the brackets is
omitted.

Volume changes are small, in transfer from liquid
to adsorbed state, so we may estimate the enthalpy
change as follows: The partial molal energy change,
in transfer from the pure liquid to the adsorbed
state, is:

$$\overline{\Delta U} = \Delta U^v - \overline{\Delta U}^{ads} \qquad (5)$$

Here, ΔU^v is the molar energy of vaporization, and
$\overline{\Delta U}^{ads}$ the partial molal energy of adsorption from the
vapor. For the purpose of a strictly Langmuir-type
computation, we assume $\overline{\Delta U}^{ads}$ to be constant.

Estimates of $\overline{\Delta U}^{ads}$ have not been brilliantly
successful in the past; but we are not directly
interested in the heat of adsorption from the gas.
Rather, we are interested in the energy of transfer
from the liquid. Thus, if we use a model that is
moderately reasonable to estimate ΔU^v, and the same
model to estimate $\overline{\Delta U}^{ads}$, we will have a good proba-
bility of making a better estimate of the difference
between these two terms. It will develop that we use
the macrospic heat of vaporization of the pure liquid
as our experimental parameter in predicting coverage.
Molecular considerations will enter only in regard to
estimating the energy of interaction between unlike
molecules or segments, and to taking molecular shape
into account. Thus, we can write, for substance i,
taking the liquid as the reference state,

$$\Delta U_i^v \cong N\epsilon_{ii}z_{iL}/2 \qquad (6)$$

Here, z_{iL} is the coordination number for substance i
in the liquid, and ϵ_{ii} is the energy at the minimum
of the pair potential function, for substance i. Let
the molecules of the liquid be designated 1, and those
of the solid 2. Then the energy change per mole, on
transfer of molecules of type 1 from the liquid to the
adsorbed state, is:

$$\overline{\Delta U} \cong N(z_{1L}\epsilon_{11}/2 - z_{1a}\epsilon_{12}/2) \qquad (7)$$

Here ϵ_{12} is the energy at the minimum of the potential function, for 1-2 bimolecular interaction, and z_{1a} is the number of nearest neighbor surface molecules, or groups, that can interact with a molecule of adsorbate. The last term the brackets in Equation 7 corresponds to the energy of adsorption from the vapor, estimated according to the same model used for Equation 6.

Equating $\overline{\Delta U}$ and $T\overline{\Delta S}$, we obtain

$$- NkT[\ln \frac{x_a}{1-x_a} + \ln(\frac{3d}{4\ell})] = N(z_{1L}\epsilon_{11} - z_{1a}\epsilon_{12})/2$$

$$= \frac{Nz_{1L}\epsilon_{11}}{2} \left[1 - \frac{z_{1a}\epsilon_{12}}{z_{1L}\epsilon_{11}} \right] \quad (8)$$

If the dominant types of attractive forces for the two species are the same, it has been shown that ϵ_{12} is given, approximately, by (10):

$$\epsilon_{12} \cong \sqrt{\epsilon_{11}\epsilon_{22}} \quad (9)$$

The right side of Equation 8 then becomes:

$$\frac{Nz_{1L}\epsilon_{11}}{2} \left[1 - \frac{z_{1a}}{z_{1L}} \sqrt{\frac{\epsilon_{22}}{\epsilon_{11}}} \right] \quad (10)$$

If the coordination number of a molecule (or group) of type 1, in the liquid, z_{1L}, is the same as that of a molecule (or group) of type 2 in its liquid state, z_{2L}, and if there is the same degree of validity for the assumption of pairwise additivity of energies for substances 1 and 2, and the same fractional contribution due to neighbors outside the first coordination shell, then,

$$\epsilon_{11}/\epsilon_{22} \cong \Delta U_1^v/\Delta U_2^v \quad (11)$$

Combining Equations 6, 10 and 11 with Equation 8, we obtain:

$$\log\frac{x_a}{1-x_a} = \frac{-\Delta U_1^v}{2.3RT} \left[1 - \frac{z_{1a}}{z_{1L}} \sqrt{\frac{\Delta U_2^v}{\Delta U_1^v}} \right] - \log(\frac{3d}{4\ell}) \quad (12)$$

A simple case, to which we can apply Equation 12, is CCl_4 on polytetrafluorethylene (Teflon TFE). We estimate the ratio of energies, $\epsilon_{22}/\epsilon_{11}$, as

$\Delta U_{CCl_4}^{V}/\Delta U_{CF_2}^{V}$. For this purpose, $\Delta U_{CF_2}^{V}$ is estimated from heat of vaporization data for CF_4 ([16]). Using the experimental energy of vaporization for CCl_4, it is found that $x_a = 3 \times 10^{-5}$.

The film pressure for a dilute monolayer can be computed from

$$\pi_e = \frac{kT}{\sigma} \frac{x_a}{1-x_a} \tag{13}$$

where σ is the area per molecule in a close-packed monolayer. For CCl_4 on Teflon, assuming $\sigma = 30A^2$, the result is $\pi_e = 4 \times 10^{-3}$ ergs/cm^2. Thus we obtain a first, important conclusion: For one liquid that forms a non-zero contact angle on a low-energy solid ([1]), π_e should be negligible, provided the solid is homogeneous and molecularly smooth.

An important system to which this theory must be applied is the series of homologous n-alkanes on polytetrafluoroethylene. For this purpose, we must modify Equations 8 to 12. We may assume the Teflon surface to consist of extended chains. The zigzag structure of the fluorocarbon chain just matches the period of zigzag in a saturated hydrocarbon chain. We may neglect the helical configuration of the fluoro-carbon, because the pitch is small, about 14 carbons for a turn of the helix. The fact that the lateral spacing between $(CF_2)_n$ chains is wider than between $(CH_2)_n$ chains does not affect this computation. For an n-alkane, the methyl groups must be treated separately from the methylenes, because the polarizability of a CH_3 is considerably greater, and a terminal CH_3 will have more nearest neighbors in the liquid than will a mid-chain CH_2. For a long-chain hydrocarbon, most of the neighbors of a CH_3 in the liquid state are CH_2 groups, so the energy of interaction between CH_3 groups in the liquid may be neglected. Then $z_{iL}\epsilon_{11}$, Equation 7, may be replaced with the expression

$$(n-2)z_{LCH_2}\epsilon_{CH_2,CH_2} + 2z_{LCH_3}\epsilon_{LCH_3,CH_2} \tag{14}$$

and $z_{1a}\epsilon_{12}$, by

$$(n-2)z_{aCH_2}\epsilon_{CH_2,CF_2} + 2z_{aCH_3}\epsilon_{CH_3,CF_2} \tag{15}$$

With this model, the expression for ΔU, employed in Equation 10, becomes

$$\Delta U = \frac{N}{2} \left[(n-2) z_{LCH_2} \epsilon_{CH_2CH_2} \left\{ 1 - \frac{z_{aCH_2}}{z_{LCH_2}} \sqrt{\frac{\epsilon_{CF_2CF_2}}{\epsilon_{CH_2CH_2}}} \right\} \right.$$

$$\left. + 2 z_{LCH_3} \sqrt{\epsilon_{CH_3CH_3} \epsilon_{CH_2CH_2}} \left\{ 1 - \frac{z_{aCH_3}}{z_{LCH_3}} \sqrt{\frac{\epsilon_{CF_2CF_2}}{\epsilon_{CH_2CH_2}}} \right\} \right] \quad (16)$$

We can estimate the ϵ's for methylene and methyl groups in terms of energies of vaporization, as was done for use in Equation 12. As before, we approximate the ratios under the square roots, by the ratio of energies of vaporization of CF_4 and CH_4. Combining the result with the entropy terms, we obtain:

$$\log \frac{x_a}{1-x_a} \cong \frac{-1}{2.3RT} \left[(n-2) \Delta U^V_{CH_2} \left\{ 1 - \frac{z_{aCH_2}}{z_{LCH_2}} \sqrt{\frac{\Delta U^V_{CF_4}}{\Delta U^V_{CH_4}}} \right\} \right.$$

$$\left. + 2 \sqrt{\Delta U^V_{CH_3} \Delta U^V_{CH_2}} \left\{ 1 - \frac{z_{aCH_3}}{z_{LCH_3}} \sqrt{\frac{\Delta U^V_{CF_4}}{\Delta U^V_{CH_4}}} \right\} \right] - \log(\frac{3d}{4\ell}) \quad (17)$$

We have carried out computations for n-alkanes on Teflon, using $\Delta U^V_{CH_2} = 3250$ joules/mole and $\Delta H^V_{CH_3} = 8200$ joules/mole, from vapor pressure data (16), and $z_{aCH_2}/z_{LCH_2} = 2/4$, $z_{aCH_3}/z_{LCH_3} = 2/7$. The ratio d/ℓ was estimated assuming the van der Waals diameter of a n-alkane to be 4.8A, and the projection of the C-C distance on the chain axis to be 1.252A. Table I shows the results of this calculation. Values for pentane and butane are in parentheses because the assumptions that most of the neighbors of a CH_3 group are CH_2 groups, and that the molecule lies flat on the surface, break down for short-chain alkanes.

We note at once, from Table I, that x_a and π_e are essentially zero for the higher hydrocarbons. Below hexane, x_a increases rapidly; and the assumption of no lateral interactions quickly becomes inapplicable. So this computation leads directly to a prediction of the region where the method of computation should break down. Thus, it

Table I.

n-alkane	x_a	π_e, ergs/cm^2
octadecane	1.4×10^{-6}	4.4×10^{-5}
hexadecane	4.2×10^{-6}	1.5×10^{-4}
decane	1.1×10^{-4}	5.6×10^{-3}
octane	4.0×10^{-4}	2.3×10^{-2}
hexane	1.7×10^{-3}	0.12
pentane	(2.9×10^{-3})	(0.23)
butane	(5.4×10^{-3})	(0.48)

is as expected, that the computed values for butane and pentane are much below those observed ([4]).

It is possible that the estimates of z_{aCH_2} and z_{aCH_3} used above may, for certain kinds of sites, be too small. For a real surface, it is highly probable that step or ledge sites exist, for which z_{aCH_2} may be 3 and z_{aCH_3} may be as 5. Such sites could well constitute 5 or 10% of an experimental surface. Examination of models shows that, for elongated molecules, it is unlikely that z_{aCH_2} could be as large as 3 for the entire chain, together with z_{aCH_2} as large as 5 for every terminal CH$_3$. Indeed the values of these two z's will depend on chain length, and on the lateral spacing of fluorocarbon chains. It is only for convenience that we approximate them by constants, and we will assume average values of 2.5 for z_{aCH_2} and 4 for z_{aCH_3}.

An alkane molecule in a site such as just described will not have freedom to take up any angular orientation parallel to the plane of the surface; indeed there will be just two orientations which have the same energy. So the term, $\log (3d/4\ell)$, in Equation 17 must be replaced by $\log (3d^2/\ell^2)$.

Table II shows the result computations of this type: The fraction $x_{a(step)}$ of step sites on Teflon TFE covered by n-alkane molecules, and the contribution to the film pressure due to this coverage, assuming 5% of the surface is accounted for as step sites.

Table II: Estimate of fractional coverage of step sites on teflon TFE by n-alkane molecules, and contribution of this coverage to equilibrium film pressure, assuming 5% of surface is accounted for as step sites.

Table II

n-alkane	$x_{a(step)}$	$0.05\pi_e$, ergs/cm^2
octadecane	8.5×10^{-4}	1.3×10^{-3}
hexadecane	1.8×10^{-3}	3.2×10^{-3}
decane	2.2×10^{-2}	5.8×10^{-2}
octane	5.2×10^{-2}	1.5×10^{-1}
hexane	1.5×10^{-1}	0.50
(pentane)	(2.0×10^{-1})	(0.8)
(butane)	(3.0×10^{-1})	(1.3)

The values for pentane and butane are given in parentheses because of the breakdown of an assumption (in addition to those noted regarding Table I) that was made in the step-site model. This is, that the average values of z_{aCH_2} and z_{aCH_3} are less than 3 and 5, respectively.

We note at once, from Table II, that the computed contributions to π_e for the higher n-alkanes is not appreciable — just as was seen in Table I. And for the lower alkanes, for which the model breaks down, the computed contribution to π_e is still relatively small — only 1.9 erg/cm^2 for butane, for which it is computed that about 30% of the step-sites are occupied.

Finally, we will treat water on Teflon and on polyethylene, using Equation 12 with omission of the term, $\log(3d/4\ell)$. To evaluate ϵ_{12}, we use the relations:

$$\epsilon_{11} = \epsilon_{11}^d + \epsilon_{11}^i + \epsilon_{11}^\mu \tag{18}$$

$$\frac{\epsilon_{11}^d}{\epsilon_{11}^d + \epsilon_{11}^i + \epsilon_{11}^\mu} = \frac{\frac{3}{4}\alpha_1^2 I_1}{\frac{3}{4}\alpha_1^2 I + \alpha_1 \mu_1^2 B + \frac{2}{3}\frac{\mu_1^4 B^2}{kT}} \tag{19}$$

$$\epsilon_{22} = \epsilon_{22}^d \tag{20}$$

$$\epsilon_{12} = \Phi_{12}^d \sqrt{\epsilon_{11}^d \epsilon_{22}} \tag{21}$$

Here Φ^d is computed by the equation

$$\Phi^d = \frac{2\sqrt{I_1 I_2}}{I_1 + I_2} \tag{22}$$

where I is ionization energy. These relations are discussed in detail in Ref. (10). The superscripts d, i and μ, in Equations 18 - 22 refer respectively to dispersion, induction and permanent dipole components of intermolecular energy. α is polarizability, I is ionization energy, μ is dipole moment, and B is a constant which, for molecules consisting of atoms in the second and third rows of the periodic table, is close to 0.66 (10). For water, the ratio computed by Equation 19 is close to 0.20 (10). Equations 18 and 19, are implicit in the discussion of Fowler and Guggenheim (17,18).

Equation 21 can be put in the form,

$$\epsilon_{12} = \epsilon_{11} \sqrt{\frac{\epsilon_{11}^d}{\epsilon_{11}} \cdot \frac{\epsilon_{22}}{\epsilon_{11}}} \tag{23}$$

$$\cong \epsilon_{11} \sqrt{0.2 \frac{\Delta U_2^v}{\Delta U_1^v}} \tag{24}$$

The reduction in average coordination number, Δz, for water, on going from bulk liquid to adsorbed state, is probably about 1, i.e. from something near 4, to about 3. Polyethylene is treated as extended $(CH_2)_n$ chains. The results are given in Table III. It should be emphasized that these computations are for water on ideal Teflon TFE and polymethylene surfaces, i.e. assuming the solids to be smooth and homogeneous.

Table III. Estimates of fractional coverage and of
equilibrium film pressure for water on a fluorocarbon
solid and on a polymetheylene surface, both assumed
to be smooth and homogeneous.

Solid	x_a	π_e, ergs/cm^2
Teflon TFE	6×10^{-7}	2.5×10^{-6}
Polyethylene	7×10^{-6}	3×10^{-5}

Discussion

(a) The Approximations. We must now make a
careful examination of the important approximations
that we have introduced. The first, and possibly the
most important, approximation is the use of the
Langmuir equation itself. The very low values of x_a
computed show that these systems (except for the
lower hydrocarbons) should be within the coverage
region where the Langmuir equation can safely be used.
We have, of course, neglected lateral interactions so
far; and we now examine the validity of that neglect.
For a quasi-spherical molecule such as CCl_4, the energy
term — i.e. the right side of Equation 12 —
becomes

$$- \frac{\Delta U_1^V}{2.3RT} \left[1 - \frac{z_{1a}}{z_{1L}} \sqrt{\frac{\Delta U_2^V}{\Delta U_1^V}} - \frac{\overline{z}_{1a}^*}{z_{1L}} \right] \qquad (25)$$

where $\overline{z}_{1a}^*$ is the average number of "lateral" nearest
neighbors of an adsorbed molecule. In a perfect
2-dimensional, hexagonal lattice, $\overline{z}_{1a}^*$ would be 6. For
CH_2 groups in an n-alkane, $\overline{z}_{1a}^*$ is at most 2. The
entropy of adsorption will be less than the Langmuir
entropy (14), so to use the expression,(25) instead
of the corresponding term in Equation 12 will yield the
maximum value of x_a:

$$\frac{x_a}{1-x_a} < \exp\left[-\frac{\Delta U_1^V}{RT} \left\{ 1 - \frac{z_{1a}}{z_{1L}} \sqrt{\frac{\Delta U_2^V}{\Delta U_1^V}} - \frac{z_{1a}^*}{z_{1L}} \right\} \right] \qquad (26)$$

For CCl_4 on Teflon TFE, with $z_{1a}^*/z_{1L} = 0.5$, this
treatment yields $x_a < 10^{-2}$. Thus, the refinement of
computing x_a with allowance for lateral interactions
does not bring the estimated coverage seriously outside the region of the isotherm where linearity is

expected. Hence, the use of the Langmuir isotherm as
an acceptable approximation, for molecules such as
CCl_4, appears to be justified. (For step sites, the
sites themselves will be isolated from each other, and
so there is no lateral interaction of adsorbate
molecules even when most of the step sites are
covered). For liquids having much lower boiling points
than CCl_4, however, it is to be expected that lateral
interactions and two-dimensional condensation will be
important, and hence that larger values of π_e will
be found.

A second approximation is the assumption of
chemical uniformity of the solid surface. It has been
established (19) that hydrophilic sites exist on at
least some, and possibly all, samples of Teflon TFE.
Such sites are, no doubt, chemically different from
the majority of sites. On polyethylene, oxygen-
containing groups are likely to be present at the
surface; and on other low-energy surfaces, it is to
be expected that high-energy heterogeneities should
commonly be present in a finite concentration. Such
sites would interact with water molecules, and could
contribute to a large value of π_e as computed from
water adsorption, yet not make an excess contribution
to hydrocarbon adsorption.

Heterogeneity will also be present for geometric
reasons, such as microscopic roughness or microporosity
beyond the level of the step-sites considered above.
So some adsorbed molecules will have more nearest
neighbors, in terms of groups such as CF_2, CF_3 etc.
on Teflon, than will others. But in general, it is
very unlikely that z_{1a} will be larger than 5 or 6.
So the ratio z_{1a}/z_{1L} will be, at most, about 0.5; and
the large values of this ratio will pertain to only
a very small fraction of surface sites, on a surface
which is smooth enough for contact angle measurements
to be made. The ratio, $\epsilon_{22}/\epsilon_{11}$, for a low-energy
solid, 2, in contact with any liquid, 1, will seldom
be very large; for a hydrocarbon on a fluorocarbon, in
the computations given above, it turned out to be
at most 1.25. There will be no more than a small
fraction of the sites for which the energy term in
Equation 12 or 17 is, for geometric reasons, very much
larger than that computed above. And for elongated
molecules on such sites, it has already been noted
that the angular entropic term (the last term in
Equations 12 and 17) will be much more negative,
because to occupy such a high-energy site, an extended
molecule will have at most two configurations accessible to it. Hence, the coverage of an essentially

flat surface will not be seriously larger than that
calculated here.

Of course, surfaces can be prepared which are
very much more heterogeneous, for geometric reasons,
than just indicated, e.g. by abrasion. But as
pointed out by Neumann and Good (20), such surfaces
are not suitable for contact angle measurements; and
the reproducibility of data obtained would be very
poor, and the hysteresis extremely large. Also, the
heterogeneity of a surface that has been modified, by
partial oxidation or by grafting short hydrophilic
chains onto it, will be much more serious than that
of an unmodified surface. The film pressure on such
surfaces is a very different problem from that on an
essentially homogeneous, low-energy surface, and it
will not be discussed here.

The term $R\ln(W_a/W_L)$ is a rigorous expression for
the change in entropy associated with angular configurations. The approximation of evaluating it as
$R\ln(3d/4\ell)$ has already been discussed. The error
thus introduced results in prediction of too high an
adsorption; so the conclusion as to the negligible
value of π_e for higher hydrocarbons is not weakened
by it.

Regarding the expressions containing the ϵ's,
e.g. Equation 10, it may be seen at once that ϵ_{11} is
by far the most important parameter, because the ratio,
z_{1a}/z_{1L} is generally small, e.g. 3/12, or rarely
larger than perhaps, 5/11. Since ϵ_{11} is evaluated
from ΔU_1^v, it is clear that the dominant energetic
component of the computation is the heat of vaporization of the liquid. There is considerably less
energy than ΔU^v "recovered", on adsorption, because
the coordination number is so much lower, for the
adsorbed molecule with the groups in the adsorbent
surface.

The estimation of ϵ_{12} from $\sqrt{\epsilon_{11}\epsilon_{22}}$, Equation 9, is
if anything, on the high side, even for systems where
the cohesion of the bulk liquid and of the solid are
both dominated by the London force. A more exact
expression, replacing Equation 9, is

$$\epsilon_{12} = \Phi \sqrt{\epsilon_{11}\epsilon_{12}} \tag{27}$$

where Φ can be computed a priori, from molecular
properties (10), e.g., by Equation 22 for nonpolar
molecules.

The fact that we express the ϵ's in terms of
ΔU^v's (e.g. Equations 11 to 17) is, in a practical

sense, a strong point of this theory. We have already noted the assumptions made — see the paragraph preceding Equation 11. The main factor leading to inequality of z_{1L} and z_{2L} (see above) is, differences in degree of expansion of the liquids, considered as fraction of quasi-lattice sites that are vacant, at the temperatures where the heats of vaporization are measured. Since the boiling point is, to a fair approximation, a "corresponding temperature" within the meaning of the theory of corresponding states, we can conclude the equality of coordination numbers is a good approximation.

Even if these errors do not cancel totally, they should do so partially. And since this ratio appears under a square root sign, Equation 12, and then is multiplied by a factor that is of the order of 0.25, the sensitivity of the theory to errors of this kind is small.

Equations 16 and 17 introduce approximations needed to treat highly asymmetric molecules. The asymmetry cannot be handled in terms of any existing, macroscopic treatments which use heats of vaporization directly. These assumptions are, however, a priori, reasonable, and so can be counted on as leading to a valid prediction of the trends.

Equations 18 to 22 are based on Good and Elbing (10) and (as already noted) on the earlier treatments of Good and Girifalco (2,7,18) and Fowler and Guggenheim (17). A notation based on that of Fowkes (21) has been employed; but the quantities are not obtainable from empirical treatment of surface tension data, as is the case with Fowkes' "γ^d". Indeed, the dispersion component of the total surface energy cannot in general be evaluated from contact angle data, because of the lack of thermodynamic uniqueness of the U^s function, in a binary system. (It is only when there is no interfacial excess mass of either component that U^s is unique. In general, $U^{s}(1) \neq U^{s}(2)$, where the superscript designates the Gibbs dividing surface: (1) refers to the surface located such that $\Gamma_1 = 0$, and (2), to the $\Gamma_2 = 0$ surface.) Thus, ϵ^d_{ii} is a function which has physical meaning only in terms of the components of intermolecular attraction constants as computed from molecular properties, e.g. by Equation 18 and 19.

The coefficient, 0.2, in Equation 24, arises from this treatment of the energy components for interaction of water and a nonpolar solid. Its use in the theory for water depends for accuracy, not so much on the validity of Equation 18 or 19, as on the assumptions about coordination number and second-nearest-neighbors, which justify the use of the ratio of energies of vaporization in Equation 24. We estimate the uncertainty here, as no worse than 50% in $\epsilon_{11}/\epsilon_{22}$, which means an uncertainty that is less than 25% after the square root is extracted. The qualitative conclusion that x_a and π_e are negligible, for H_2O on Teflon and polyethylene, are not seriously changed by allowance for such uncertainty, because ΔU^v is so large, for water, and as a consequence, the computed values of x_a and π_e are so small.

In summary, the approximations of this theory are probably not serious; and indeed, the majority of the approximations contribute errors in the direction of too high an estimate of x_a and π_e. It would take very much greater correction factors than now seem likely to appear, in any refined theory, to change the qualitative conclusion that x_a and π_e are generally small for liquids that boil above room temperature.

It must be emphasized, however, that these computations are not intended as <u>quantitative</u> predictions of x_a and π_e for these systems. The approximations made above, including those about structure of the solid surface, are sufficiently serious that quantitative agreement cannot be expected. But <u>for the prediction of trends</u>, as with the n-alkanes, and for the conclusion that π_e is small for high boiling liquids, the approximations should not detract from the validity of this theory.

(b) <u>Comparison with Experiments</u>: π_e for Organic Compounds on Teflon.

Graham (<u>4</u>) has found π_e to be much smaller, for n-octane on powdered Teflon TFE, than for alkanes having surface tension below γ_c, e.g. butane. π_e has been reported to be large for low-boiling gases such as N_2 on Teflon. Graham (<u>4</u>) noted a difficulty arising from bulk solubility of an alkane in the substrate, such that the quantity sorbed could not be uniquely assigned as between adsorption and absorption. (This trouble was not encountered by Whalen and Wade (<u>5</u>,<u>6</u>)). Graham concluded that his value of π_e for octane on Teflon, about 1.7 erg/cm^2, was an upper limit; he could not estimate the real value. Wade and Whalen (<u>5</u>,<u>6</u>) also used a powder form of Teflon, and also obtained a small value for π_e: 3.3 for hexane and 2.9 for octane. They explicitly corrected for pendular ring condensation, and in so doing, they obtained estimates of π_e in their systems that were very probably more valid than Graham's were for his. Our computations are lower than these experimental results

e.g. π_e = 0.1 for hexane in the absence of step sites.
The result, π_e = 0.5 for Teflon with step sites
accounting for 5% of the area, shows the direction of
change in the computed results when a more complex
surface is postulated. The preliminary computations
made for CCl_4 indicate that the explicit employment
of a two-dimensional van der Waals equation (11,12)
would be worthwhile with the lower hydrocarbons. We
will investigate such two-dimensional condensation in
another communication. We note, however, that our
equations have successfully predicted the observed
trend of π_e with chain length.

As has already been noted, water adsorption
measurements at Lehigh (19) have shown that the surface
of Teflon powder is heterogeneous, and about 0.75% of
the surface of the sample studied consisted of sites
which adsorbed water strongly. The computed adsorption of water, reported in Table III, is for the 99%
of the surface exclusive of the hydrophilic sites.
The Lehigh measurements (19) could not quantitatively
distinguish the increment of water adsorption due to
this fraction of the surface. The water isotherm
was described as resembling a type II isotherm, with
surface area less than 1% of the nitrogen area,
superimposed on a type III isotherm. We have not
attempted to analyse the contribution to the adsorption
of hydrocarbons or other nonpolar molecules, due to
the hydrophilic sites on Teflon. Wade (6) has
discussed the evidence for interactions of high-energy
sites on Teflon with hydrocarbon molecules. Whalen
(22) has recently measured the adsorption of water on
Teflon powder, and estimated a value of 1.9 ergs/cm^2
for π_e. He estimates the hydrophilic sites on this
solid to account for 3% of the surface area (6,23).

Adsorption measurements on Teflon powders are of
limited direct pertinence to the contact angle question
because there is every reason to suspect that the
surface of a powder is quite significantly different
from that of a smooth slab that is suitable for
contact angle measurement. This point is hard to
verify directly, however, because while electron
microscopic examination of a grossly smooth solid
could reveal pores, it would not reveal chemical
heterogeneity, particularly that in the form of isolated, high-energy sites; and the precision measurement of contact angle on a powder is not easy.

(c) Comparison With Adamson's Model, and His
Results For Water On Polyethylene.

(1) <u>Theory</u>. Our model is, in principle, not incompatible with that of Adamson and Ling ($\underline{3}$), in that their isotherm graph, in its lower pressure region, resembles a BET type II isotherm. It is well known that, by lowering the value of the "c" constant, a type III isotherm is obtained which, in its low-coverage region, is not experimentally distinguishable from the Henry's law region of a Langmuir isotherm.

Figure 1 shows a schematic graph of the isotherm we propose. Since the coverage at saturation is far below a monolayer, there is no need to make any hypotheses about the structure of the absorbed material in the region above monolayer coverage.

(2) <u>Experiment</u>. Adamson, in 1973, reported ($\underline{8},\underline{9}$) a value of 42 ergs/cm^2 for π_e of water on polyethylene. These results appear to be in disagreement with Table III. They are also in apparent conflict with the well-known conclusion ($\underline{24}$) that the γ_c polyethylene is in the neighborhood of 31 ergs/cm^2. The use of Equation 28,

$$\gamma_S = \frac{[\gamma_L(\cos\theta+1) - \pi_e]^2}{4\Phi^2\gamma_L}$$

$$\simeq \frac{\gamma_L(\cos\theta+1)^2}{4\Phi^2}$$

(28)

with Φ about 0.9 to 0.95 ($\underline{2},\underline{10}$) leads to an estimate of about 36 ergs/cm^2 for γ_S. It would be rather anomalous for polyethylene in water vapor to have a negative surface free energy γ_{SV}, which would be obtained by subtracting 42 from 36 as Equation (1) would seem to prescribe.

A reconciliation can be achieved by hypothesizing that water on the polyethylene sample that was used ($\underline{8},\underline{9}$) behaved in a similar fashion to water on Teflon TFE, ($\underline{19}$) noted above. The water molecules might adsorb in clusters on isolated hydrophilic sites, yielding a film with the observed average thickness and the reported π_e. <u>With respect to water adsorption</u>, the <u>effective</u> γ_S of this solid would then be, not the value derived from contact angles of nonpolar liquids, but a larger value:

$$\gamma_S^* = \gamma_S + \pi_{e(water)}$$

$$\sim 36 + 42$$

$$= 78 \text{ ergs/cm}^2$$

(29)

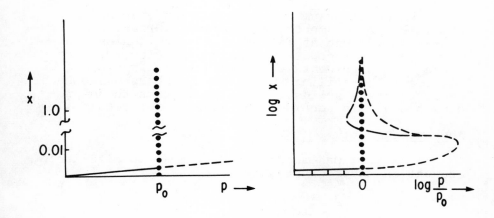

Figure 1. Schematic: A possible isotherm for adsorption of a liquid such as carbon tetrachloride or hexadecane on a molecularly smooth, homogeneous solid such as Teflon TFE. Extrapolations shown on log-log plot are speculation, and no physical reality is intended, particularly beyond the region where the curve starts to swing to the left.

46   ADSORPTION AT INTERFACES

If the water clusters on Adamson's polyethylene were isolated from each other, their presence would not reduce the water contact angle to zero. And in the hydrophobic regions between clusters, the amount of water adsorbed per unit area, and π_e, would be negligible — as computed above, for Table III.

Conclusions

The equilibrium film pressure, π_e, should be negligible on a smooth, homogeneous surface of a low-energy solid such as Teflon, for most liquids that form non-zero contact angles, and particularly for those with high heats of vaporization. π_e should be large for low-boiling substances, such as N_2, ethane, etc., on these solids. It is shown that these predictions are in accord with experimental results.

Acknowledgement

This work was supported by the National Science Foundation under Grant GK10602.

Literature Cited

1. Fox, H.W. and Zisman, W.A., J. Colloid Sci., (1950) 5, 514.

2. Good, R.J. and Girifalco, L.A., J. Phys. Chem., (1960) 64, 561.

3. Adamson, A.W., and Ling, I., in "Contact Angle, Wettability and Adhesion", p. 57, Advan. Chem. Ser No. 43, American Chemical Society, Washington, D.C., 1964.

4. Graham, D.P., J. Phys. Chem., (1965), 69, 4387.

5. Wade, W.H., and Whalen, J.W., J. Phys. Chem. (1968), 72, 2898.

6. Whalen, J.W., J. Colloid and Interface Sci., (1968), 28, 443.

7. Good, R.J., in "Contact Angle, Wettability and Adhesion", Advan. Chem. Ser. No. 43, American Chemical Society, p. 74, Washington, D.C., 1964.

8. Adamson, A.W., 47th National Colloid Symposium, Ottawa, Canada, June, 1973.

9.   Adamson, A.W., 167th National Meeting of the
      American Chemical Society, Los Angeles, April, 1974.

10.   Good, R.J. and Elbing, E., Ind. Eng. Chem. (1970),
      62, (3), 54.

11.   de Boer, J.H., "The Dynamical Character of Adsorp-
      tion," Oxford University Press, London, 1953.

12.   Ross, S., and Olivier, J.P., "On Physical Adsorp-
      tion," Interscience Publishers, New York, 1964.

13.   Hill, T.L., "Introduction to Statistical Thermo-
      dynamics," Addison-Wesley Publishing Co., Reading,
      Mass., 1960.

14.   Everett, D.H., Proc. Chem. Soc., Feb., 1958, p. 57.

15.   Glasstone, S., Laidler, K.J., and Eyring, H.,
      "Theory of Rate Processes," McGraw-Hill, N.Y.,1941.

16.   C.R.C. Handbook, 51st ed., Chemical Rubber
      Publishing Co., Cleveland, Ohio, 1970.

17.   Fowler, R.H., and Guggenheim, E.A., "Statistical
      Thermodynamics," Cambridge University Press,
      London, 1952.

18.   Berghausen, P.E., Good, R.J., Kraus, G., Podolsky,
      B., and Soller, W., "Fundamental Studies of the
      Adhesion of Ice to Solids," WADC Technical
      Report 55-44, 1955.

19.   Chessick, J.J., Healey, F.H., and Zettlemoyer,
      A.C., J. Phys. Chem. (1956), 60, 1345.

20.   Neumann, A.W. and Good, R.J., J. Colloid and Inter-
      face Sci., (1972), 38, 341.

21.   Fowkes, F.M., J. Phys. Chem. (1957), 61, 904.

22.   Whalen, J.W., Vacuum Microbalance Techniques, A.W.
      Czanderna, Ed., v. 8, p. 121, Plenum Press, 1971.

23.   Whalen, J.W., (personal communication, 1974).

24.   Zisman, W.A., in "Contact Angle, Wettability and
      Adhesion," Advan. Chem. Ser., No. 43, pp. 1-51,
      American Chemical Society, Washington, D.C., 1964.

4

Binding of Solute and Solvent at the Interface and the Gibbs Surface Excess

D. K. CHATTORAJ and S. P. MOULIK

Department of Food Technology and Biochemical Engineering and Department of Chemistry, Jadavpur University, Calcutta-32, India

Introduction

With the help of an imaginary mathematical plane placed at the interfacial region, a liquid-gas system, according to the Gibbs concept, may be divided into two phases. From the thermo-dynamic analysis of such a system, Gibbs also derived his well-known equation relating the surface excess of solute with the surface tension and bulk activity of the solute in solution. The physical concepts associated with the Gibbs surface excess have been examined more critically by Guggenheim and Adam (1) and also by Defay and Prigogine (2). Guggenheim (3) has also given an alternative derivation of the Gibbs adsorption equation assuming certain arbitrary but finite values for the physical thickness of the interfacial phase. In some cases, the inter-facial thickness may be estimated from the experimental data (4). Goodrich (5) has recently analyzed the Gibbs adsorption equation with the help of an algebric method in which no mention is made of the mathematical dividing surfaces. Surface excesses of solute and solvent in a two-component system are, however, relative to each other and their values may be positive or negative. An attempt will be made in this paper to obtain re-lations between the surface excesses and the absolute amounts of solvent and solute bound to the interfacial layer.

It is well-known that the surface tension of water in-creases with addition of various inorganic electrolytes to its bulk. In the dilute region of electrolyte concentrations, this increase is explained in terms of image forces (6,7,8). The extensive application of the Gibbs adsorption equation for the highly concentrated electrolyte solutions did not attract many workers because interpretation of the negative surface excess of electrolyte in terms of its interfacial adsorption is diffi-cult. An attempt will also be made here to estimate the compo-sition of the interfacial phase of an electrolyte solution from the observed negative value of the solute surface excess.

Binding of Two Components at the Liquid Interface

    Let us imagine that a certain amount of a liquid component
1 is mixed with a definite amount of another liquid or solid
component 2 as a result of which a binary solution in the shape
of a column (Figure 1) is formed. With the help of a plane set
arbitrarily at position p_1p_1', this column may be divided into
two distinct regions A and B. The exact nature and the characteristic feature of this dividing plane will be discussed later
on in detail. Region B, composed solely of the bulk liquid
phase, contains n_1 and n_2 moles of components 1 and 2 respectively. The components in the bulk phase are said to remain in
the free state of mixing. In region A, n_1' and n_2' moles of two
components are present as a whole. Out of these, n_1' and n_2''
moles existing in the free state of mixing may form the bulk
phase. Remaining $\Delta n_1'$ moles of component 1 are believed to mix
with $\Delta n_2'$ moles of component 2 in an entirely different manner
because of the existence of the interfacial energy at the upper
extreme top side of this region. They are termed to remain in
the bound state of mixing and considered to form one square
centimeter of an interface.

    Let Γ_1 and Γ_2 be the relative excesses of components 1
and 2 respectively in region A which may be defined by the relations (2),

$$\Gamma_1 = n_1' - n_2' \frac{X_1}{X_2} \qquad \ldots \qquad (1)$$

$$\Gamma_2 = n_2' - n_1' \frac{X_2}{X_1} \qquad \ldots \qquad (2)$$

where X_1 and X_2 are bulk mole fractions of the respective components in region B. The concentrations of solute and solvent
in the gas phase are neglected in writing Equations (1) and (2).
Γ_1 and Γ_2 will be zero when the composition n_1'/n_2' of region
A becomes equal to the bulk composition X_1/X_2 of region B.
Combining relations (1) and (2), Equation (3) will be obtained.

$$X_1 \Gamma_2 + X_2 \Gamma_1 = 0 \qquad \ldots \qquad (3)$$

    For given values of X_1 and X_2, the values of Γ_1 and Γ_2
are thus related to each other according to Equation (3) and
obviously the two excesses are opposite in sign. Further, at
definte values of X_1 and X_2, n_1' and n_2' in Equations (1) and
(2) are not fixed but are significantly dependent on the arbitrary position of the dividing plane. The absolute composition
of the surface cannot therefore be expressed in terms of Γ_1,
Γ_2, n_1' and n_2'.

Let us now assume further that the dividing plane at p_1p_1' behaves as a semi-permeable membrane which allows only the components in their free state of mixing to pass from region A to region B and vice versa. However, the components in their bound state of mixing in region A are not allowed to permeate through this membrane. The chemical potential μ_2'' of component 2 in region A, remaining in the free state may be given by the relation,

$$\mu_2'' = \mu_2^o + RT \ln f_2 \left[\frac{n_2''}{n_1'' + n_2''} \right]$$

$$= \mu_2^o + RT \ln f_2 \left[\frac{n_2' - \Delta n_2'}{(n_1' - \Delta n_1') + (n_2' - \Delta n_2')} \right] \quad \ldots \quad (4)$$

Here f_2 and μ_2^o are the rational activity coefficient and standard chemical potential of component 2 in the free phase respectively. The chemical potential of the same component in region B is similarly given by Equation (5).

$$\mu_2 = \mu_2^o + RT \ln f_2 X_2 \quad \ldots \quad (5)$$

From the principle of membrane equilibrium, μ_2 is equal to μ_2'' so that combining Equations (4) and (5), one will obtain

$$\frac{n_2' - \Delta n_2'}{n_1' - \Delta n_1'} = \frac{X_2}{X_1} \quad \ldots \quad (6)$$

Rearranging this,

$$n_2' - n_1' \frac{X_2}{X_1} = \Delta n_2' - \frac{X_2}{X_1} \Delta n_1' \quad \ldots \quad (7)$$

Equation (7) is similar in form to that used by Bull and Breese (9) for the calculations of the extents of solute and solvent binding on the protein boundary. Inserting Equation (7) in Equation (2),

$$\Gamma_2 = \Delta n_2' - \Delta n_1' \frac{X_2}{X_1} \qquad \ldots \quad (8)$$

It may similarly be shown that

$$\Gamma_1 = \Delta n_1' - \Delta n_2' \frac{X_1}{X_2} \qquad \ldots \quad (9)$$

For given values of X_1 and X_2, $\Delta n_1'$ and $\Delta n_2'$ (and hence Γ_1 and Γ_2) are not dependent on the position of the dividing membrane. Absolute composition of the bound phase at the interface may be obtained if $\Delta n_1'$ and $\Delta n_2'$ are evaluated from the experimental data.

Equation (2) may also be written in the form,

$$\Gamma_2 = \left(\Delta n_2' - \Delta n_1' \frac{X_2}{X_1} \right) + \left(n_2'' - n_1'' \frac{X_2}{X_1} \right) \quad \ldots \quad (10)$$

From the comparison relations (8) and (10), one will at once find,

$$\frac{n_2''}{n_1''} = \frac{X_2}{X_1} \qquad \ldots \quad (10a)$$

This means that the mole ration of the components in the free phases of regions A and B are identical. In the right side of Equation (2), two equal and opposite variables n_2'' and $n_1'' X_2/X_1$ become implicitly included owing to the arbitrary placement of the dividing plane including a part of the bulk region. The numerical values of these variables then depend upon the position of this dividing plane. The terms in relations (8) and (9) are independent of the position of the dividing membrane.

The Gibbs Adsorption Equation

Let us now imagine that the dividing plane is raised from position $p_1 p_1'$ to $p_2 p_2'$ slowly and reversibly, so that all the free components from region A permeate to region B. Region B is now occupying the total bulk of the liquid system whereas region A, solely composed of $\Delta n_1'$ and $\Delta n_2'$ moles of the bound components, may be regarded to form the surface phase of one square centimeter interfacial area. The special mixing of the components at the interface is obviously due to the existence

of the interfacial energy. In the absence of the imaginary
membrane also, the bound state should remain as a physical
reality because the interfacial free energy does not disappear
under this condition. The surface bound phase may thus be
imagined to be separated entirely from the liquid in bulk, by
an imaginary phase-dividing plane placed at the position P_2P_2'.
Applying the Gibbs-Duhem relations for the surface (bound) and
bulk phases separated with such a plane set appropriately, we
can obtain the familiar expressions for the Gibbs adsorption
equation which at constant temperature and pressure will read
(3),

$$\Gamma_2 = (\Delta n_2' - \Delta n_1' \frac{X_2}{X_1})$$

$$= \frac{f_2 X_2}{RT} \frac{d\gamma}{d f_2 X_2} \qquad \dots \quad (11)$$

and

$$\Gamma_1 = (\Delta n_1' - \Delta n_2' \frac{X_1}{X_2})$$

$$= -\frac{f_1 X_1}{RT} \frac{d\gamma}{d X_1 f_1} \qquad \dots \quad (12)$$

Here γ stands for the surface tension of the liquid mixture
corresponding to a given bulk mole fraction X_2 of the solute,
f_1 and f_2, the rational activity coefficients of the solvent
and solute respectively, R the gas constant and T the absolute
temperature. In the conventional forms of the Gibbs equation,
Γ_1 and Γ_2 are related to n_1' and n_2' according to Equations
(1) and (2) respectively which on their turn depend significantly on the position of an arbitrary dividing plane p_1p_1'.
     Δn_1 and Δn_2 are, however, independent of the position of
the dividing plane.

Evaluations of Γ_2

     The surface tension, γ, of water usually increases with
increase in concentration of an inorganic electrolyte provided
the solute content in the aqueous solution is high. For the
present analysis, the data on the surface tension of water for
different high concentrations of an electrolyte have been collected form the literature. These electrolytes are the following: LiCl (10) 25°C; NaCl (11) 25°C; KCl (12) 25°C; NaBr (10)

25°C; NaNO$_3$ (10) 20°C; K$_2$SO$_4$ (12) 25°C; MgCl$_2$ (10) 20°C; MgSO$_4$ (13) 25°C and Al$_2$ (SO$_4$)$_3$ (10) 25°C. The surface tension data of non-ionic sucrose (10) solutions have also been considered here since this solute is observed to increase the boundary tension of water. The practical activity coefficients (on the molality scale) for various concentrations of these electrolytes and sucrose can be obtained from the literature (14). The values of the corresponding rational activity coefficients (f_2) of these solutions may easily be calculated using appropriate conversion factors (14). Here water and solute stand for components 1 and 2 respectively. For each solute, γ has been plotted against f_2X_2 and the slopes dγ /d (f_2X_2) of the curve for various values of f_2X_2 have been calculated using the chord-area method of slope analysis (15). The negative surface excesses -Γ_2, of the electrolytes, for various values of X_2, have been calculated with the help of Equation (11). In Figures 2, 3, 4 and 5, - Γ_2 for various electrolytes and sucrose is plotted against the mole ratio X_2/X_1. The open and closed signs in the figures represent data obtained from experiments and from interpolation of γ-f_2X_2 curves, respectively.

Solute-Solvent Binding Characteristics

It is of interest to note that - Γ_2 in all these plots in Figures 2, 3, 4 and 5 increases linearly with increasing mole ratio, X_2/X_1, provided the solute content in the solution is not too high. The slope and intercept of such linear plots may evidently represent the respective values of $\Delta n_1'$ and $\Delta n_2'$ according to Equations (8) or (11). The least square values of $\Delta n_1'$ and $\Delta n_2'$ for various systems are presented in Table I along with their respective standard deviations.

From the values of $\Delta n_1'$ in Table I, it may be observed that one square centimeter of a solution interface may bind significant amount of water whose magnitude may be of the order 10^{-8} to 10^{-9} moles. $\Delta n_1'$ for alkali chlorides follow the order: LiCl < NaCl < KCl. The water binding capacities of the interface in the presence of uni-univalent sodium salts (NaCl, NaBr and NaNO$_3$) do not differ from each other widely. The magnitudes of $\Delta n_1'$ for uni-univalent and uni-bivalent salts (K$_2$SO$_4$, MgCl$_2$) are also comparable to each other. However, $\Delta n_1'$ for polyvalent electrolytes, Al$_2$ (SO$_4$)$_3$ and MgSO$_4$, are significantly high. The agreement of the order of $\Delta n_1'$ for various electrolytes with that to be expected from the lyotropic series is only partial.

From Figures 2, 3, 4 and 5, it may also be noted that the plot of- Γ_2 against X_2/X_1 deviates from linearity for LiCl, NaBr (not shown in figure), NaNO$_3$, K$_2$SO$_4$ (not shown), and Al$_2$ (SO$_4$)$_3$ when the mole ration X_2/X_1 are relatively high. This may indicate that the surface bound components $\Delta n_1'$ and $\Delta n_2'$ are not constant but become variable functions of X_2/X_1 at relatively higher solute concentrations. When X_2/X_1 for NaNO$_3$, K$_2$SO$_4$, KCl,

Table I. Least Square Values of $\Delta n'_1$ and $\Delta n'_2$

Serial No.	Electrolytes	$\Delta n'_1 \times 10^9$ moles/cm^2	$\Delta n'_2 \times 10^{11}$ moles/cm^2	L_s	$\dfrac{\Delta n'_2}{\Delta n'_1}$	m_s
1	LiCl	2.95 ± 0.31	0.34 ± 1.04	1.77		
2	NaCl	3.70 ± 0.07	-0.15 ± 0.12	2.21		
3 (a)	KCl (dil)	5.05 ± 0.01	-1.31 ± 0.39	3.02		
(b)	KCl (conc)	5.57 ± 0.18	-1.51 ± 0.79	3.33		
4	NaBr	3.06 ± 0.05	0.01 ± 0.08	1.83		
5 (a)	NaNO$_3$ (dil)	3.28 ± 0.07	-0.39 ± 0.08	1.97		
(b)	NaNO$_3$ (conc)	8.36 ± 0.70	57.7 ± 9.24	3.45	0.069	3.90
6 (a)	K$_2$SO$_4$ (dil)	4.71 ± 0.96	-0.08 ± 0.12	2.82		
(b)	K$_2$SO$_4$ (conc)	3.74 ± 0.04	-0.01 ± 0.03	2.24		
7	MgCl$_2$	6.66 ± 0.29	0.51 ± 0.23	3.99		
8 (a)	MgSO$_4$ (dil)	52.7 ± 0.90	0.10 ± 0.35	31.6		
(b)	MgSO$_4$ (conc)	155 ± 0.83	-0.05 ± 0.61	93.0		

Table I. (Contd.)

Serial No.	Electrolytes	$\Delta n_1' \times 10^9$ moles/cm²	$\Delta n_2' \times 10^{11}$ moles/cm²	L_s	$\dfrac{\Delta n_2'}{\Delta n_1'}$	m_s
9(a)	Al$_2$(SO$_4$)$_3$ * (dil)	32.0	0	19.2		
(b)	Al$_2$(SO$_4$)$_3$ (conc)	17.6 ± 2.92	4.66 ± 4.23	10.5	0.0026	0.14
10(a)	Sucrose (dil)	4.34 ± 1.73	2.62 ± 1.70	2.60	0.0060	0.33
(b)	Sucrose (conc)	15.1 ± 0.89	18.8 ± 1.86	9.00	0.0124	0.69

*not least square values.

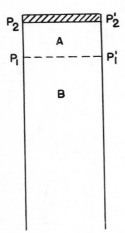

Figure 1.   *Hypothetical binary solution in the shape of a column*

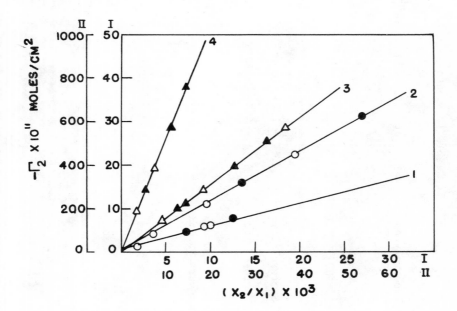

Figure 2.   *Surface excess of solute as a function of mole ratio. 1: KCl (Scale I-I). 2: KCl (Scale I-II). 3: $MgSO_4$ (Scale II-II). 4: $MgSO_4$ (Scale I-I).*

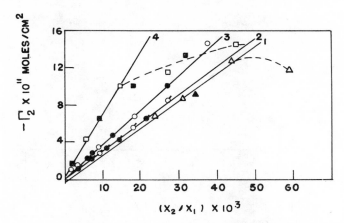

Figure 3.   Surface excess of solute as a function of mole ratio.
1–4: LiCl, NaBr, NaCl, MgCl$_2$.

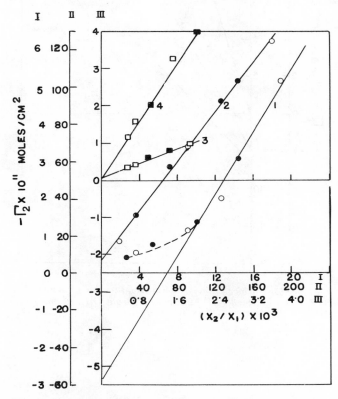

Figure 4.   Surface excess of solute as a function of mole ratio.
1: NaNO$_3$ (Scale II-II). 2: NaNO$_3$ (Scale I-I). 3: K$_2$SO$_4$ (Scale
III-III). 4: K$_2$SO$_4$ (Scale III-I).

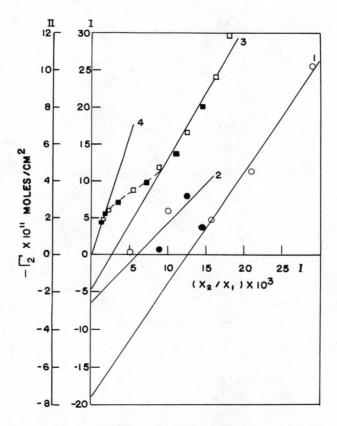

Figure 5.   Surface excess of solute as a function of mole ratio.
1: Sucrose (Scale I-I).   2: Sucrose (Scale II-I).   3 and 4:
$Al_2(SO_4)_3$ (Scale I-I).

$MgSO_4$, $Al_2 (SO_4)_3$ and sucrose are very high, $-\Gamma_2$ vs X_2/X_1
plots become again linear having different slopes and intercepts. The least square values of $\Delta n_1'$ and $\Delta n_2'$, calculated
again from Equation (11) at these regions of high solute content, are also included in Table I.

Compared to $\Delta n_1'$, values of $\Delta n_2'$ for different electrolytes are indeed quite small. $\Delta n_2'$ of $NaNO_3$ (concentrated) is
very high and the surface molality (m_s) in this case is as high
as 4 molal compared to its bulk molality varying between 5 to
10. For $Al_2 (SO_4)_3$ and sucrose, $\Delta n_2'$ is significant and m_s lies
between 0.10 to 0.70. For all other cases, $\Delta n_2'$ and m_s are less
than 10^{-11} and 10^{-2} respectively. The observed experimental
error for the evaluation of $\Delta n_2'$ are quite high so that the
signs and magnitudes of $\Delta n_2'$ for these electrolytes have no
physical meaning. For these electrolytes, one may easily neglect $\Delta n_2'$ in Equation (11) without making any significant error
in the evaluation of $\Delta n_1'$. In the exceptional case of KCl,
$\Delta n_2'$ is of the order 10^{-11}, and its signs are positive both in
the high and low ranges of mole ratio. $\Delta n_1'$ for low and
high KCl concentrations are found to be 5.0×10^{-9} and 5.6×10^{-9}
moles per unit area respectively. Since negative values for
$\Delta n_2'$ do not bear any physical meaning, probably $\Delta n_2'$ in
Equation (11) for KCl is also negligible but $\Delta n_1'$ instead of
remaining constant slowly increases with gradual increase of
X_2/X_1. The initial slope representing $\Delta n_1'$ of KCl in a short
range of electrolyte concentration will not be, however, too
far off from the observed value 5.0×10^{-9}. In general, we
conclude that only a small amount of electrolyte is associated
with the surface-bound water even when a large amount of the
same electrolyte remains dissolved in the bulk water phase.
Interfacial water for these cases is highly surface-active and
it has a strong dislike for inorganic electrolytes and sucrose.

The relative positive surface excess of water in the conventional interpretation of the Gibbs equation may be identified
with the amount of water adsorbed on one square centimeter of
the liquid interface when the adsorbed solute is arbitrarily
taken as zero. In many of the cases considered here, $\Delta n_2'$ is in
fact negligibly small so that according to Equations (3), (8)
and (9), Γ_1 is very close to $\Delta n_1'$. Results in Table I for
$NaNO_3$ and a few electrolytes also indicate that $\Delta n_1'$ may be
significantly greater thann Γ_1 if $\Delta n_2'$ is not negligibly small.
Further, the relative excess Γ_2 which is negative is observed
to increase considerably with increase of X_2/X_1. Under these
conditions, electrolyte actually bound, $\Delta n_2'$, is either negligibly small or significantly positive and at certain range of
electrolyte concentration, it becomes independent of the mole
ratio X_2/X_1.

Assuming the effective cross-sectional area of a water
molecule to be 1 nm^2 at the interface, the moles (Δn_1^*) of
water required to form one square centimeter (6) of an inter-

facial monolayer is 1.7×10^{-9}. Neglecting the contribution of
$\Delta n_2'$, the number (L_s) of water layers in the bound interfacial
region may be estimated from the ratio $\Delta n_1' / \Delta n_1^*$. L_s for
various electrolytes and sucrose are also included in Table I.
The results indicate that the surface bound water is multi-
molecular for all solutes. For uni-univalent and uni-bivalent
electrolytes, as well as for non-ionic sucrose, the number of
such surface bound layers may lie between the values 2 to 6.
Values of L_s, for $Al_2(SO_4)_3$ and sucrose (concentrated) are
moderately high whereas those for $MgSO_4$ in both high and low
concentration ranges are unusually large. Higher values of L_s
for polyvalent electrolytes may be real. It may alternatively
be possible that these polyvalent electrolytes undergo considerable hydrolysis or ion-association at high solute concentrations.
Under these conditions, several types of solute components will
exist in the system so that significant corrections may be
needed for the calculation of the surface excess with the help
of the Gibbs adsorption equation (16).

Conclusions

It is evident from all these discussions that a liquid
column in Figure 1 may be divided by a phase demarcating plane
P_2P_2' such that every part of the bulk phase below this plane
has the mole ratio composition X_2/X_1. Above this plane, the
mole ratio of any part of the surface-bound liquid phase may
deviate from X_2/X_1 because of the binding of $\Delta n_1'$ and $\Delta n_2'$
moles of the components. Similar real concept for the surface
phase has also been discussed by others (2, 3). With the help
of the Gibbs Equations (11) and (12), the actual composition of
the surface region in terms of $\Delta n_1'$ and $\Delta n_2'$ may be evaluated
from the surface tension-concentration data. This evaluation is
only possible when the composition of the surface bound liquid
phase becomes independent of the bulk mole ratio so that $- \Gamma_2$
becomes a linear function of X_2/X_1.

The liquid column in Figure 1 may further be divided by
placing a semipermeable membrane at an arbitrary position P_1P_1'
so that region A now contains the bulk and the surface-bound
phases respectively below and above the phase dividing plane
P_2P_2'. The relative surface excesss Γ_2 may thus be shown to
be equal to either of the expressions occurring on the left
or right side of Equation (7). The magnitudes of the terms
n_1' and n_2' in this equation depend significantly on the arbitrary position of P_1P_1' whereas those of $\Delta n_1'$ and $\Delta n_2'$ are invariant to P_1P_1'.

In the conventional derivation of the Gibbs adsorption
equation, sometimes the phase-dividing plane is arbitrarily
shifted from its fixed position P_2P_2' to a variable position
P_1P_1'. The whole region of A is then arbitrarily imagined to be
the surface phase composed of n_1' and n_2' moles of the components.

The individual values of n_1' and n_2' will not bear any physical signifance. However, the relative surface excess Γ_2 defined in terms of n_1' and n_2' according to Equation (2) may be correctly calculated from the $\gamma - X_2$ data since the relative surface excess does not depend on the position of P_1P_1'. The invariant character of the relative excess may be clear from the close scrutiny of Equations (7), (10) and (10a).

Unlike the long-chain ions, the cations and anions of the inorganic electrolytes are not specially adsorbed and oriented at the interface of water. Both surface and bulk phases divided by the plane P_2P_2' should be individually electroneutral (2) in terms of the net charge of cations and anions. With the help of the Gibbs Equation (11), one can only calculate the macroscopic composition of the surface phase in terms of the water and electrolyte components bound to each other. The microscopic feature of the ionic double layer and the structure of water and voids in the interfacial region cannot be predicted from the present analysis.

From the values of L_S in Table 1, it may be concluded that the interfaces of different electrolyte solutions usually possess physical thickness of the order 1 to 10 nm. This order is in agreement with that suggested by Guggenheim (17) for the thickness of the physically defined surface phase. In contrast to the uniform composition of the bulk region, the solute and solvent bound surface region should be physically inhomogeneous (17, 18). Assuming this physical picture of the surface and bulk phases, Guggenheim (17, 18) has deduced the Gibbs adsorption equation which is similar in form to Equations (11) and (12). However, the individual amounts of the solute and solvent components bound at the interface in his treatment remain arbitrary and uncertain since the bulk phase is separated from surface phase by arbitrary placement of a dividing plane. From the discussions presented above, it is noted that $\Delta n_1'$ and $\Delta n_2'$ are independent of the imaginary placement of the dividing plane. Their values are absolute and under certain conditions may be evaluated within the limits of experimental error. In a separate paper it will be shown that the present approach can also be useful to calculate the interfacial binding of components resulting from the mixing of an organic liquid with water.

Acknowledgement

The authors are grateful to Dr. K. K. Kundu for many helpful discussions.

Literature Cited

1. Guggenheim, E.A. and Adam, N.K., Proc. Roy. Soc. London, Ser. A, (1933) 139, 231.

2. Defay, R., Prigogine, I., Belleman, A. and Everett, D.H.,
    "Surface Tension and Adsorption", John Wiley and Sons, Inc.,
    New York, 1966.
3. Guggenheim, E.A., Trans. Faraday Soc., (1940), 36, 397.
4. Chattoraj, D.K. and Chatterjee, A.K., J. Colloid Interface
    Sci., (1966) 21, 159.
5. Goodrich, F.C., Trans. Faraday Soc., (1968) 64, 3403.
6. Randles, J.E.B., in "Advances in Electrochemistry and
    Electrochemical Engineering", P. Delahay, Ed., Vol. 3, p.1,
    Interscience, New York 1963.
7. Onsager, L. and Samaras, N.N.T., J. Chem. Phys., (1934) 2,
    528.
8. Buff, F.P. and Stilbinger, F.H., Jr., Chem. Phys., (1956) 25,
    312.
9. Bull, H.B. and Breese, K., Arch. Biophys. Biochem., (1970)
    137, 299.
10. "Hand Book of Chemistry and Physics", 47th ed., F25,
    The Chemical Rubber Company, Cleveland, Ohio, 1966-67.
11. Jones, G. and Ray, W.A., J. Amer. Chem. Soc., (1941) 63,
    3262.
12. Jones, G. and Ray, W.A., J. Amer. Chem. Soc., (1937) 59,
    187.
13. Jones, G. and Ray, W.A., J. Amer. Chem. Soc., (1942) 64
    2744.
14. Robinson, R.A. and Stokes, R.H., "Electrolyte Solutions",
    2nd ed. (revised), Butterworts, London, 1959.
15. Chatterjee, A.K. and Chattoraj, D.K., J. Colloid Interface
    Sci., (1968) 26, 140.
16. Chattoraj, D.K. and Pal, R.P., Indian J. Chem., (1970) 10,
    417.
17. Guggenheim, E.A., "Thermodynamics", North Holland, Amsterdam,
    1961.
18. Adam, N.K., "The Physics and Chemistry of Surfaces", Dover
    Publications, New York, 1968.

Mechanism of Sulfonate Adsorption at the Silver Iodide–Solution Interface

K. OSSEO-ASARE, D. W. FUERSTENAU, and R. H. OTTEWILL*

Department of Materials Science and Engineering, University of California, Berkeley, Calif. 94720

Introduction

The early work on silver iodide dispersions ([1]) showed that the sols prepared in the presence of an excess of iodide ions were more stable than those prepared with silver ions present in excess. This was found to be due to the fact that the point of zero charge did not coincide with the equivalence point, i.e. in the presence of potential determining ions only, the equivalence point was found to be at pAg 8 whereas the net charge on the surface became zero at pAg 5.5. Historically silver iodide has been classified as a hydrophobic sol, a definition largely based on the fact that silver iodide sols are not reversible, i.e. coagulated sols cannot be redispersed by diluting with water (see Frens and Overbeek ([2])). Recently, studies of the adsorption of water vapor on silver iodide powders have shown that the surface is also hydrophobic in terms of the usual definition of wettability. Zettlemoyer et al. ([3]) found that the surface area ratios water/argon and water/nitrogen were much less than unity and concluded that approximately three out of five surface sites were hydrophobic; they attributed the hydrophilic-hydrophobic balance to oxide impurities.

The results of Pravdic and Mirnik ([4]) showed that at pI 5 (pAg 11) and pH 5 hexylamine can reverse the charge of a silver iodide particle. This behavior is in marked contrast to the effect of alkylamine on quartz, a hydrophilic substrate, where as far as adsorption is concerned the octylammonium ion appears to behave in the same way as an ammonium ion ([5]). Long chain alkyl sulfates, even at low concentrations, also affect the zeta-potential of the silver iodide surface ([6]). Moreover, according to Bijsterbosch and Lyklema ([7]) even short chain alcohols, e.g. n-butyl, adsorb with their hydrophobic parts towards the silver iodide. Recently measurements of the contact angle of silver iodide surfaces in water have shown that values of ca. 23° can be obtained ([8,9]). These factors all suggest that there is strong hydrophobic interaction between hydrocarbon chains and

silver iodide surfaces which can play an important role in silver
iodide-surface active agent solution interactions.

In addition, recent studies using electron microscopy have
shown that alkyl pyridinium compounds appear to undergo a chemical reaction with silver iodide surfaces at surfactant concentrations approaching the critical micelle concentration (10).
It appears therefore that the adsorption of surface active agents
at a silver iodide surface may involve a complex interplay of
electrical, hydrophobic (dispersion forces) and chemical interactions.

The purpose of the present paper is to elucidate adsorption
mechanisms in the alkyl sulfonate-silver iodide system through
electrophoretic mobility measurements. The sulfonates used in
this study ranged from pentyl to tetradecyl. Further information
on the nature of the AgI surface and sulfonate adsorption was
obtained through study of contact angle phenomena in these
systems.

Experimental - Materials and Methods

Materials. Triply distilled water used for all solutions
was prepared by first passing tap water through a Barnstead
Laboratory still followed by a two-stage Haraeus quartz still.
The final distillate was kept under purified nitrogen until used.

Silver nitrate and potassium iodide of A.R. grade were used
to prepare the silver iodide sol. Iodine was obtained as resub-
limed crystals from Mallinckrodt and silver wire of 99.5 to 99.8%
purity was obtained from Sargent-Welch. The sodium salts of C_{14},
C_{12}, C_{10}, and C_8 sulfonic acids were supplied by Aldrich Chemical
Company while the C_5 was from K & K Laboratories.

Preparation of Silver Iodide Sol. A stock sol was prepared
by adding 50 ml of 10^{-2} M KI solution to an equal volume of 10^{-2}
M $AgNO_3$ solution with stirring. After aging for 12 - 18 hours,
this was diluted to a sol concentration of 10^{-4} M AgI for electrophoresis measurements. The ionic strength was controlled with
10^{-3}M KNO_3 and appropriate volumes of $AgNO_3$ and surfactant solutions were added to give the required pAg and surfactant concentration.

Electrophoresis. The electrophoretic measurements were conducted with the Riddick Zetameter (11), a product of Zetameter
Inc., New York. Room temperature was maintained at 20 ± 2°C for
most of these experiments. In general, twenty mobility readings
were taken for each system and averaged; the polarity of the
applied voltage was reversed for alternate measurements. Usually
100 V was used; however, for slower particles it was often necessary to go to voltages as high as 300 V to observe any appreciable motion. Care was taken not to keep the sol too long before
use since on standing for extended periods of time, the sol

particles have been found to decrease in mobility - probably a
result of the desorption which accompanies coagulation (6).

The average diameter of the sol particles, 2a, was found by
electron microscopy (12) to be 20 nm. The ionic strength was
controlled using 10^{-3} M KNO_3; this gave a Debye-Huckel reciprocal
length $1/\kappa$ of 9.5 nm. Based on the fact that under these conditions the product κa has a value of 1.05, the results of Wiersema
et al. (13) can be used to convert the measured electrophoretic
mobilities to zeta potentials. Below a mobility, u_e, of 2 μm/
sec per volt/cm, as a good fit to the calculations of Wiersema
et al., the mobility can be converted into zeta potential by

$$\zeta = 20u_e \tag{1}$$

where ζ is the zeta potential in millivolts. At the higher mobility values, however, the relationship between mobility and
zeta potential becomes nonlinear.

Wetting Behavior.

Preparation of thin AgI films. Following the method of
Billett and Ottewill (8), glass microscope slides, cut into
pieces approximately 1 cm x 2.5 cm x 0.1 cm, were cleaned with
aqua regia and stored under triply distilled water. To plate
the glass with silver, a watch glass with the slides in it, was
positioned 3 cm vertically below a silver source (approximately
6 cm silver wire) held in a tungsten basket. The vacuum unit
was pumped down to about 666.6 μPa through a liquid nitrogen trap.
The silver mirrors were transferred quickly from the vacuum evap-
orator and immersed in a 0.0025 N solution of iodine in 0.01 M KI
solution for 20 - 30 seconds. The films were then aged in 10^{-4}
M KI solution for 1.5 hours and kept under distilled water until
used. It was found that leaving a thin layer of silver between
the glass plate and the AgI film improved the adhesion of the
film to the substrate. This procedure was therefore followed in
the preparation of the films.

Measurement of contact angle. The captive bubble technique
(8) was used to determine the contact angles. For making the
measurement, the solution to be studied was first added to a
cell made of optical glass and then the slide with the AgI film
was placed into the desired solution for about 15 minutes before
taking measurements. A glass cylindrical tube of 3 mm internal
diameter was placed above the film surface and the air pressure
in the bubble was regulated by means of a screw arrangement at
the upper end of the tube. A telescope supplied with an ocular
protractor was used to observe the bubble profile.

The bubble was allowed to touch the film surface by gently
increasing the pressure, then after sitting on the film for about
15 seconds, the pressure was increased by a small amount. This

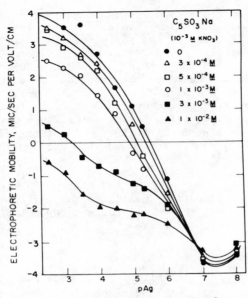

*Figure 1. The effect of pAg on the electro-
phoretic mobility of silver iodide in the presence of sodium pentyl sulfonate at millimolar
ionic strength with potassium nitrate*

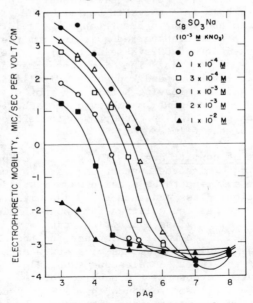

*Figure 2. The effect of pAg on the electro-
phoretic mobility of silver iodide in the presence of sodium octyl sulfonate at millimolar
ionic strength with potassium nitrate*

situation yields the receding angle θ_R. The pressure was then
released until the contact boundary on the plate just moved.
The resulting angle is the advancing angle, θ_A. Surfactant solutions of a given concentration were prepared and the contact
angles were measured as a function of pAg for C_{10} and C_{14}. In
all cases, including the experiments in the absence of surfactant,
the ionic strength was controlled with 10^{-3} M KNO_3. The contact
angles reported here are the average of six or more values obtained by placing the bubbles on different samples or on different positions on the same sample.

Results

Electrophoresis. Figures 1 to 5 present the electrophoretic
mobility of AgI as a function of pAg for various alkyl sulfonate
concentrations and different chain lengths. In the absence of
surfactant, the point of zero charge of AgI was found to occur
at pAg 5.6, in agreement with previous values reported in the
literature (1). In all cases, with increasing concentrations of
surfactant at any pAg values below the pzc, the positive mobility
of the sol particles decreased, passed through zero, and then
became increasingly negative.
The most interesting features of these results are
(1) The short chain sulfonates are able to reverse the zeta
potential on AgI although much higher concentrations are required
than for longer chain sulfonates.
(2) All the curves appear to coincide in the neighborhood
of pAg 7 independent of chain length or surfactant concentration.
(3) The zeta potential goes through a maximum at higher
concentrations of surfactant, an effect which increases with
chain length.

Wetting Behavior. The magnitude of the contact angles obtained both in the absence of surfactant and in the presence of
various concentrations of C_{14} and C_{10} have been plotted as a
function of pAg in Figures 6 to 8.
In the absence of surfactant, both the receding and advancing contact angles go through a maximum at about pAg 5.4. The
curves shown in Figure 6 are drawn through the weighted average
of the experimentally determined contact angle values for each
pAg. The spread in the θ values for a given pAg is largely the
result of the inherent difficulty of obtaining perfectly reproducible surfaces. These results are in good agreement with those
reported by Billett and Ottewill (14) who also observed a maximum
at about pAg 5.4, which is close to the pzc.
Figures 7 and 8 show that when the bulk concentration of the
surfactant is low, the wetting behavior is very similar to that
of the surfactant-free system except at higher positive surface
charge (low pAg) where a significant rise in contact angle is
observed. In other words, in Figure 8, the receding contact

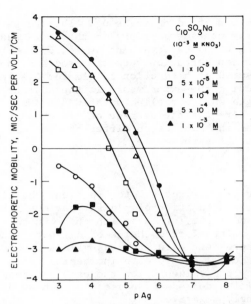

*Figure 3. The effect of pAg on the electro-
phoretic mobility of silver iodide in the presence of sodium decyl sulfonate at millimolar
ionic strength with potassium nitrate*

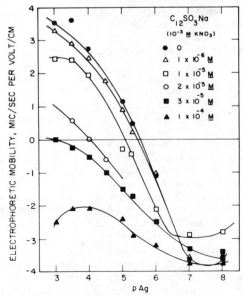

*Figure 4. The effect of pAg on the electro-
phoretic mobility of silver iodide in the presence of sodium dodecyl sulfonate at millimolar
ionic strength with potassium nitrate*

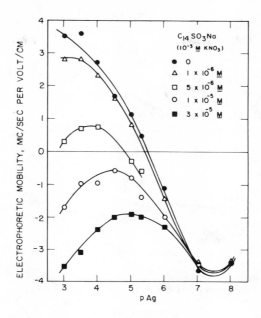

*Figure 5. The effect of pAg on the electro-
phoretic mobility of silver iodide in the presence of sodium tetradecyl sulfonate at millimolar ionic strength with potassium nitrate*

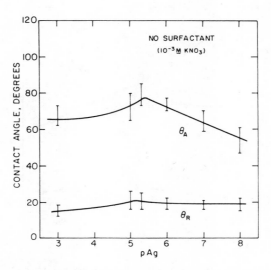

*Figure 6. The contact angle on silver iodide in
the absence of surfactant as a function of pAg at
millimolar ionic strength with potassium nitrate*

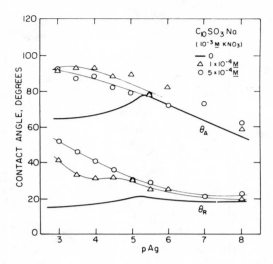

Figure 7. The contact angle on silver iodide in the presence of sodium decyl sulfonate as a function of pAg at millimolar ionic strength with potassium nitrate

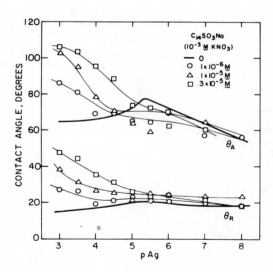

Figure 8. The contact angle on silver iodide in the presence of sodium tetradecyl sulfonate as a function of pAg at millimolar ionic strength with potassium nitrate

angles in the presence of 10^{-6} M C_{14} sulfonate show the same
variation with pAg as that found for the surfactant-free system.
However, at pAg values lower than pAg 4, there is a rise in the
contact angle values; e.g. for pAg 3 in 10^{-6} M C_{14} sulfonate,
θ_R = 27° compared with a value of 18° for the surfactant-free
system. As the concentration of surfactant in the bulk solution
increases, a corresponding increase in the contact angle is also
observed, e.g. for pAg 3 in 3 x 10^{-5} M C_{14} sulfonate, θ_R = 47°
while it is 27° in 10^{-6} M solution. However, at all surfactant
concentrations, it appears that for pAg values above the pzc,
there is very little influence of surfactant on the contact
angles, i.e. there is relatively less sulfonate adsorption.

Discussion

The results obtained indicate, in agreement with previous
workers, that alkyl sulfonates are strongly adsorbed on silver
iodide surfaces when the latter are positively charged. Adsorption also appears to occur at the point of zero charge and at low
negative charge densities on the surface. As the magnitude of
the negative charge increases, however, the electrokinetic results obtained become indistinguishable from those obtained in
the absence of the surfactant, suggesting that if adsorption does
occur, it is only in small quantities and it occurs in such a way
as not to influence the zeta-potential. The contact angle measurements clearly substantiate the fact that silver iodide is a
hydrophobic solid and that this must be taken into consideration
in analyzing adsorption behavior in these systems.
Interpretation of these results can be achieved through
application of the Stern–Grahame theory for adsorption in the
Stern layer (15,16). The adsorption density of surfactant ions
in the Stern layer, Γ_δ, in moles per cm^2 is given by

$$\Gamma_\delta = \frac{\Gamma_m}{1 + \frac{1}{x}\exp(\Delta G_{ads}/RT)} \qquad (2)$$

where Γ_m is the adsorption density at monolayer coverage, x is
the mole fraction of the surface-active agent in solution, and
ΔG_{ads} is the electrochemical standard free energy of adsorption.
Application of this expression involves the implicit assumption
that the size of the adsorbing species is very much smaller than
the radius of the adsorbent particle and that lateral interaction
between the adsorbing species is negligible. At low surface
coverage, Equation (2) simplifies to the Grahame equation

$$\Gamma_\delta = 2rc \exp(-\Delta G_{ads}/RT) \qquad (3)$$

where r is the effective radius of the ionic head and c is the bulk concentration of the adsorbed ions in mol/cm^3.

In general, ΔG_{ads} can be considered as the sum of electro-static and non-electrostatic terms,

$$\Delta G_{ads} = \Delta G_{el} + \Delta G_{sp} \tag{4}$$

where ΔG_{el} is the free energy change due to electrostatic inter-actions experienced by the adsorbing species at the Stern plane ($= zF\psi_\delta$ if z is the valence and F the Faraday constant), ΔG_{sp} is the free energy change due to specific interactions at the surface and in this system may include the following terms: $\Delta G_{CH_2}^{**}$, $\Delta G_{CH_2}^{*}$ and ΔG_{chem}. $\Delta G_{CH_2}^{**}$ represents the free energy change due to van der Waals association of the hydrocarbon chains with each other forming hemimicelles; $\Delta G_{CH_2}^{*}$ is the free energy associated with the van der Waals interaction of the hydrocarbon chains with the solid surface; and ΔG_{chem} is the free energy change contributed by chemical bonding of the adsorbing species with the solid surface. The process of adsorption is termed "physical" if the electrostatic and van der Waals interactions constitute the driving force for adsorption. When on the other hand, the adsorbing species form chemical bonds with ions or atoms in the solid surface, the process is termed "chemisorption" (17,18).

In order to ascertain the role which specific processes play in the adsorption of alkyl sulfonates on silver iodide, the mag-nitude of the specific adsorption free energy must be evaluated. Two methods can be used (18), based respectively on considerations of the properties of the Stern layer: (1) at the condition where the surfactant reverses the electrophoretic mobility (i.e. the point of zeta potential reversal, or the pzr) and (2) at the con-dition where the mobility -pAg curve for the surfactant system meets the indifferent electrolyte curve.

Evaluation of ΔG_{ads} at the pzr. The condition for charge balance in the electrical double layer is given by,

$$\sigma_o + \sigma_1 + \sigma_2 = 0 \tag{5}$$

where σ_o = surface charge density
σ_1 = charge density of the Stern layer
σ_2 = charge density of the diffuse double layer
Rewriting Equation 3 in the form of surface charge density gives,

$$\sigma_1 = 2rczF \exp(-\Delta G_{ads}/RT) \tag{6}$$

Moreover, the surface charge density in the diffuse double layer can be written as,

$$\sigma_2 = \sqrt{\frac{2c\epsilon RT}{\pi}} \sinh\left(\frac{zF\psi_\delta}{2RT}\right) \tag{7}$$

where ϵ is the dielectric constant in the solution. The capacity of the Stern layer, C_δ, is given by

$$C_\delta = \frac{d\sigma_0}{d(\psi_0 + \psi_\delta)} \tag{8}$$

where ψ_0 is the total double layer potential. For the condition that $\psi_\delta = 0$, then $\sigma_2 = 0$, giving,

$$\sigma_0 = -\sigma_1 \text{ and } C_\delta = \frac{\sigma_0}{\psi_0} = -\frac{\sigma_1}{\psi_0} \tag{9}$$

if the Stern layer capacity is constant. Furthermore ΔG_{el} must be zero and the total free energy of adsorption will be made up solely of the ΔG_{sp} terms. Hence,

$$\psi_0 C_\delta = -\sigma_1 = -2r\,c_0\,zF\exp(-\Delta G_{sp}/RT) \tag{10}$$

where c_0 is the bulk concentration of surfactant in mol cm^{-3} at which ψ_δ becomes equal to zero.

In Stern's theory of the electrical double layer, C_δ is assumed to be constant. By using this assumption, Grahame calculated values for the differential capacity of the double layer between mercury and aqueous solutions of salts for various concentrations of electrolyte which were in reasonably good agreement with experimental results. Thus, for the present calculations, the constancy of C_δ will be assumed and the value of 15μ F/cm^2 shall be used (19). The effective radius of the sulfonate ion is taken as 0.29 nm and ψ_0 is evaluated using the Nernst equation

$$\psi_0 = \frac{RT}{F} \ln\left(\frac{a_{Ag+}}{a^\circ_{Ag+}}\right) = -\frac{RT}{F} \ln\left(\frac{a_{I^-}}{a^\circ_{I^-}}\right) \tag{11}$$

where a°_{Ag+} and $a^\circ_{I^-}$ are the activities of Ag+ and I$^-$ respectively at the pzc.

Table I summarizes the results obtained for ΔG_{sp} using Equation (10) for pAg 3, 4, and 5, respectively. There is a strong dependence on chain length, the values of ΔG_{sp} increasing from 4.7 RT for C_5 to 10.9 RT for C_{14}. The effect of pAg does not seem to be significant for the short chain sulfonates, e.g.

Table I.   Calculation of ΔG_{sp} with Equation (10)

for the Adsorption of Alkyl Sulfonates on Silver Iodide

$(C_m = 15\mu F/cm^2,\ r = 0.29\ nm)$

pAg	Alkyl Sulfonate	$c_o \times 10^3$ mol cm^{-3}	$-\Delta G_{sp}$
	C_5	3.5×10^{-3}	4.7 RT
	C_8	2.5×10^{-3}	5.0 RT
3	C_{10}	1.3×10^{-4}	8.0 RT
	C_{12}	3.2×10^{-5}	9.4 RT
	C_{14}	7.0×10^{-6}	10.9 RT
	C_5	2.2×10^{-3}	4.7 RT
	C_8	1.5×10^{-3}	5.0 RT
4	C_{10}	7.5×10^{-5}	8.0 RT
	C_{12}	2.0×10^{-5}	9.4 RT
	C_{14}	5.6×10^{-6}	10.6 RT
	C_5	8.0×10^{-4}	4.6 RT
	C_8	1.5×10^{-4}	6.2 RT
5	C_{10}	2.3×10^{-5}	8.1 RT
	C_{12}	1.1×10^{-5}	8.9 RT
	C_{14}	3.4×10^{-6}	10.0 RT

C_5 has a constant value of 4.7 RT for the three pAg values considered. C_{10} maintains the same value of ΔG_{sp} = 8.0 RT for pAg 3 to 5.   The values for the longer chain sulfonates, however, appear to show some pAg dependency.   Thus at pAg 3 and 4, ΔG_{sp} for C_{12} is 9.4 RT, but this decreases to 8.9 RT at pAg 5.   This is in reasonable agreement with the results of Ottewill and Watanabe (6) who found ΔG_{sp} = 8.8 RT at pAg 3 for C_{12} sulfonate.   Similarly for C_{14}, ΔG_{sp} decreases from pAg 3 to 5 in the order 10.9 RT, 10.6 RT, and 10.0 RT.

Evaluation of ΔG_{ads} at the point of coincidence of mobility-pAg curves.   The second condition for evaluating the specific adsorption free energy, i.e. utilizing the pAg at which all the mobility − pAg curves coincide, provides a different means of analyzing adsorption phenomena.   At this point, it seems that the AgI surface charge is so highly negative that the consequent

Table II. Calculation of ΔG_{sp} for Hydrophobic Interaction (pAg 5)

Surface Active Agent	$c_o \times 10^3$	$-\Delta G_{sp}$ from Equation (10)	$-\Delta G_{sp} - 3.1\ RT$	$\frac{\text{Chainlength}}{2}$
C_5	8.0×10^{-4}	4.6 RT	1.5 RT	2.5
C_8	1.5×10^{-4}	6.2 RT	3.1 RT	4
C_{10}	2.3×10^{-5}	8.1 RT	5.0 RT	5
C_{12}	1.1×10^{-5}	8.9 RT	5.8 RT	6
C_{14}	3.4×10^{-6}	10.0 RT	6.9 RT	7

large ΔG_{el} almost completely opposes specific adsorption. This means that at this point, the specific adsorption potential should just be counterbalanced by the electrostatic contribution to the adsorption free energy. Thus, under this condition,

$$\Delta G_{sp} \approx -\Delta G_{el} = -zF\psi_\delta \qquad (12)$$

Since all the curves come together by pAg 7, independent of chain length and concentration of surfactant, $\psi_\delta \simeq -78$ mV when specific adsorption ceases. This yields a value of 3.1 RT for the specific adsorption free energy. In addition, because this adsorption phenomenon does not exhibit any chain length dependency, it cannot be attributed to the chain-chain and chain-solid interactions. It is therefore suggested that this behavior is caused by the specificity of the polar head of the surfactant. In particular, the interaction of the sulfonate head with Ag+ in the lattice gives rise to a chemisorption bond and this latter calculation estimates its magnitude, i.e., $\Delta G_{chem} = -3.1$ RT. Beekley and Taylor (20) have reported a marked specific influence of certain anions, including $C_{10}H_7O_3^-$, in the adsorption of Ag+ on AgI. The recent work of Ottewill and co-workers (10) also implies that there might be chemisorption in the adsorption of dodecyl pyridinium bromide on AgI. Therefore, taking the ΔG_{sp} value calculated with Equation (12) as $\Delta G_{chem} = -3.1$ RT, the hydrophobic contribution to the total specific adsorption free energy can be calculated by subtracting $\Delta G_{chem} = -3.1$ RT from the ΔG_{sp} values given by Equation (10). These have been tabulated in Table II.

There is a striking similarity between the values of $(-\Delta G_{sp} - 3.1\ RT)$ and the numbers obtained by dividing the respective chain lengths by 2, especially for the longer chain surfactants.

The free energy decrease accompanying the complete removal of a hydrocarbon chain from water is about 1 RT per mole of CH_2 groups (17). Thus, on this basis, if all the chains are removed from water, the decrease in free energy will be 5 RT, 8 RT, 10 RT, 12 RT, and 14 RT respectively for C_5, C_8, C_{10}, C_{12}, and C_{14}.

If on the other hand, the adsorbed surfactant consists of horizontally oriented chains, then the accompanying decrease in free energy might be 1/2 RT per mole of CH_2 groups (instead of 1 RT) since even though half of the chain surface is next to the solid surface, the other half is still exposed to the water. On this basis therefore, the free energy associated with complete horizontal orientation of the chains will be 2.5 RT, 4 RT, 5 RT, and 7 RT respectively for C_5, C_8, C_{10}, C_{12}, and C_{14}. On comparing these values with the corresponding $(-\Delta G_{sp} - 3.1 \text{ RT})$ values (Table II) viz. 1.5 RT, 3.1 RT, 5.0 RT, 5.8 RT, 6.9 RT respectively for C_5, C_8, C_{10}, C_{12}, and C_{14}, it seems reasonable to conclude that the adsorbed sulfonate ions orient their chains parallel to the solid surface indicating the presence of chain-solid hydrophobic interactions $(\Delta G^*_{CH_2} = \Delta G_{sp} + 3.1 \text{ RT})$. The fact that in comparison with the other chain lengths, C_5 and C_8 have values of $\Delta G^*_{CH_2}$ which are below their respective nRT/2 (where n is the number of carbon atoms in the chain) values shows that C_5, and C_8 are less surface active than their longer chain counterparts.

At higher sulfonate concentrations and especially with the C_{14} sulfonate (Figure 5), the slope of the mobility – pAg curves is reversed as the pAg is reduced. This reversal probably indicates the onset of hemimicelle formation, which is the condition where the adsorption free energy becomes more negative because of the association of the hydrocarbon chains of the adsorbed sulfonate ions.

Contact angle behavior. In view of the fact that water seems to be only weakly bonded to the AgI surface (3,21,22), there will be a strong tendency for incoming sulfonate ions to dislodge water molecules at the surface. The result will be an interface consisting of CH_2 groups with a consequent increase in the hydrophobic nature of the solid-solution interface. This means that in the presence of adsorbed sulfonates the contact angle at the solid-liquid-air interface should increase with the number of CH_2 groups at the solid surface, i.e. with both chain length and amount of adsorbed ions. Figures 7 and 8 clearly support the above analysis. A comparison of the contact angles for C_{10} and C_{14} shows that at pAg 3 a receding angle of about 50 degrees is achieved in 5×10^{-4} M C_{10} sulfonate but only 3×10^{-5} M is necessary to attain the same contact angle with C_{14} sulfonate. Again, for each surfactant concentration, the angles increase with decreasing pAg, showing that the amount of surfactant adsorbed increases with higher positive surface charge. When the pAg is low, the solid surface has a high positive charge and this leads to an increase in electrostatic attraction of the anionic surfactants to the AgI surface. This means that more surfactant ions will be drawn towards the surface and consequently the contributions of chain-solid interactions to the free energy of adsorption will be expected to increase with increased surface charge, and the longer the chain length, the more this increase will be apparent, both in electrokinetic behavior and in contact angle behavior.

Summary

The effect of n-alkyl sulfonates on the electrophoretic and wetting behavior of silver iodide has been studied. The results show that in the region of negative charge, all the surfactants cease to be surface active at the same pAg, in the neighborhood of pAg 7. This fact has been used to postulate a chemical contribution of the polar head to the total specific free energy of adsorption. It is demonstrated by means of the Stern-Grahame theory of specific adsorption at the double layer that the sulfonates adsorb at lower coverages with their chains horizontally oriented to the AgI surface. The free energy change associated with this specific chain-solid interaction is found to be 0.5 RT per mole of CH_2 group. At higher adsorption densities, and especially for longer chain lengths, association contributes to the adsorption process.

Acknowledgements

The support of this research by the National Science Foundation is gratefully acknowledged.

Literature Cited

1. Kruyt, H. R., Ed., "Colloid Science", Vol. I, Elsevier, Amsterdam, 1952.
2. Frens, G., Overbeek, J. Th. G., J. Colloid Interface Sci., (1971), <u>36</u>, 286
3. Zettlemoyer, A. C., Tcheurekdjian, N., Chessick, J. J., Nature (London), (1961), <u>192</u>, 653
4. Pravdic, V., Mirnik, M., Croat. Chem. Acta, (1960), <u>32</u>, p. 1.
5. Fuerstenau, D. W., J. Phys. Chem.,(1956), <u>60</u>, 981
6. Ottewill, R. H., Watanabe, A., Kolloid-Z., (1960), <u>170</u>, 132
7. Bijsterbosch, B. H., Lyklema, J., J. Colloid Sci., (1965), <u>20</u>, 665
8. Billett, D. F., Ottewill, R. H., in "Wetting", S. C. I. Monograph No. 25, p. 253, Society of Chemical Industry, London, 1967.
9. Billett, D. F., Hough, D. B., Lovell, V., Ottewill, R. H., (1973), unpublished data.
10. Billett, D. F., Ottewill, R. H., Thompson, D. W., in "Particle Growth in Suspensions", S. C. I. Monograph No. 38, p. 195, Academic Press, London, 1973.
11. Zeta-Meter Manual, Zeta-Meter Inc., New York, N.Y.

12. Horne, R. W., Ottewill, R. H., J. Photo. Sci., (1958), <u>6</u>, 39

13. Wiersema, P., Loeb, A., Overbeek, J. Th. G., J. Colloid Interface Sci., (1966), <u>22</u>, 78

14. Billett, D. F., Ottewill, R.H., (1971), 161st A. C. S. National Meeting, Los Angeles, Ca.

15. Stern, O., Z. Elektrochem., (1924), <u>30</u>, 508

16. Grahame, D. C., Chem. Rev. (1947), <u>41</u>, 441

17. Fuerstenau, D. W., Journal of the International Union of Pure and Applied Chemistry, (1970), <u>24</u>, 135

18. Osseo-Asare, K., Fuerstenau, D. W., Croat. Chem. Acta, (1973), <u>45</u>, 149

19. Mackor, E., L., Recl. Trav. Chim., (1951), <u>70</u>, 763

20. Beekley, J. S., Taylor, H. S., J. Phys. Chem., (1925), <u>29</u>, 942

21. Tcheurekdjian, N., Zettlemoyer, A. C., Chessick, J. J., J. Phys. Chem., (1964), <u>68</u>, 773

22. Hall, P. G., Tomkins, F. C., Trans. Farad. Soc., (1962), <u>58</u>, 1734

*Visiting Professor, Department of Materials Science and Engineering, University of California, Berkeley; Permanent address: School of Chemistry, University of Bristol, Bristol BS8 1TS, England

Adsorption of Dyes on Water-Swollen and Non-Swelling Solid Substrates

S. R. SIVARAJA IYER and A. S. CHANEKAR
Bombay University, Department of Chemical Technology, Bombay, India
G. SRINIVASAN
Hindustan Lever Research Centre, Bombay, India

Introduction

Extensive investigations carried out by Sivaraja Iyer and coworkers (1,2,3) on adsorption of anionic dyes on cellulose surfaces revealed that the extent of dye uptake was strongly influenced by the concentration and nature of the electrolyte ionic species added to the dyebath. An important effect observed (3) was a marked influence of the nature of the electrolyte cation on dye uptake in the presence of equivalent concentrations (0.1M) of different alkali metal chlorides, the order of increasing dye adsorption being Li < Na < K < Rb < Cs. An examination of the usual factors in the cellulose - aqueous dye solution system which are influenced by the addition of electrolytes such as the Donnan potential (4,5,6,7), Donnan distribution of ions (8,9,10), increase in chemical potential of the dye anion (11,12) and changes in solubility of the dye (3) shows that these factors for the series of the cations studied cannot explain the marked difference in dye adsorption in the presence of Li^+ and Na^+ on one hand and K^+ Rb^+ and Cs^+ on the other. This additional influence of the nature of the cations however, finds a reasonable explanation in terms of their varying abilities to disrupt the structure of ordered water molecules surrounding the cellulose surface and the adsorbing dye anions (3). In the present work, these studies have been extended to cover a wider range of concentrations of KCl and NaCl and also the influence of different halide anions on the extent of dye adsorption.

It is well known that cellulose is a hydrophilic material having a negative surface charge due to ionization of intrinsic carboxyl groups and when immersed in water, swells up considerably, exposing a large internal surface, also hydrophilic in nature. The influence of this opening up of its structure on dye and electrolyte sorption, water uptake etc. has been extensively reviewed (12,13,14). The question arises whether the above mentioned effect of electrolyte cations on the adsorption of dyes is confined to a swelling hydrophilic textile fiber such as viscose, or is a phenomenon associated with hydrophilic surfaces in general. Hence studies on the adsorption of the anionic dye Chlorazol Sky Blue FF on other hydrophilic substrates such as rigid amorphous titanium dioxide and amorphous silica having thickening and thixotropic character have been undertaken. It was also considered important to compare the dye adsorption characteristics of a completely hydrophobic surface like that of Graphon and a surface exhibiting dual characteristics such as activated charcoal.

The spectrum of surfaces investigated in this work was found to exhibit interesting characteristics and responses to the presence of electrolytes in the aqueous dyebath. The results of these comparative investigations are discussed in this paper.

Experimental

Preparation, Purification and Quality of Materials. The anionic direct dye Chlorazol Sky Blue FF (C.I. Direct Blue 1, sodium salt) used in these studies, was obtained in a highly pure form (purity > 99.2%) from a commerical sample using the standard salting out method of Robinson and Mills (15) for purification. This dye will be referred to in the subsequent portions of this paper as CSBFF.

The viscose rayon sample used in the present investigations was a Gwalior Rayon sample having a denier of 1.5 and was purified using standard methods (16). The specific surface area of this viscose sample as determined from a B.E.T. nitrogen adsorption isotherm for a water swollen uncollapsed sample (2) as well as by negative sorption of chloride ions in a viscose-aqueous KCl solution system (17) was found to be 205 m^2/g. The amorphous titanium dioxide (TiO_2) sample obtained through the courtesy of Professor R.D. Vold, University of Southern California, was a National Lead sample having code No. MP 1391-1 and a specific surface area of 100 m^2/g. The Graphon was a pure sample obtained through the courtesy of Professor A.C. Zettlemoyer, Lehigh University Bethlehem Pa, and has a corrected specific surface area of 120 m^2/g (18). Graphon provides a homogeneous and hydrophobic nonporous surface and is therefore widely used in adsorption studies. The silica (SiO_2) sample under the trade name Aerosil 200 was obtained as a gift from M/S. Degussa, Frankfurt, West Germany and has a specific surface area of about 200 m^2/g. The activated charcoal was microporous E. Merck (No. 2184) sample and was further purified by treating it with dilute hydrochloric acid at 70°C for 1 hr. to remove acid soluble impurities followed by repeated washings with conductivity water. This sample was then dried at 100°C. The specific surface area of the sample was 800 m^2/g. The acidity of surface groups was about 0.8 equiv. per kg.

The surface areas of all these adsorbents were obtained from B.E.T. nitrogen adsorption isotherms.

Measurement of Dye Adsorption. Known weights of the solids
(0.1 to 0.35 g depending on the specific surface area of sample)
were placed in Corning glass bulbs which were degassed under a
vacuum of 10^{-6}mm of Hg for a few hours at 100°C and then sealed.
These sealed bulbs containing the solids were broken under known
volumes of dye solutions in standard ground joint stoppered Corning glass conical flasks. The flasks and their contents were
equilibrated in a thermosated water bath having an accuracy of
$\pm$ 0.1°C. A suitable arrangement was made for agitating the flasks
during the period required for equilibrum. The equilibrium times
of adsorption for the different systems were found to be about
150 hr. for viscose, 100 hr. for TiO_2 and SiO_2, 60 hrs. for
Graphon and about 120 hr. for activated charcoal. Stock solutions
of dye and electrolyte were prepared in double distilled conductivity water and stored in thoroughly cleaned, steamed and dried
Corning conical flasks and diluted as required. Estimations of
the original and equilibrium concentrations of CSBFF in the dyebath were made spectrophotometrically with a Hilger Spekker instrument using Beer's law. The colorimetric measurements were
carried out at a wavelength of 625 nm which is the wavelength of
maxiumum absorption.

Results and Discussion

Adsorption of CSBFF on Viscose. The adsorption isotherms
determined at 50°C in the presence of increasing concentrations
of NaCl and KCl are illustrated in Figure 1. Increase in the
extent of dye uptake with increasing electrolyte concentration
clearly noticed in this figure is an expected and previously reported behaviour (2,3,12). More interesting is the observation
that at any given equivalent electrolyte concentration in the
range 0.045-0.2M the presence of K^+ ions induces much greater dye
adsorption on viscose than Na^+ions. These results amply confirm
and extend the results on dye uptake reported previously (3) at
one equivalent concentration, namely, 0.1M. Whereas cations are
seen to strongly influence the dye uptake by viscose, anions are
reported to have very little influence (12). Equilibrium adsorption of CSBFF on viscose at 50°C carried out in the presence of
0.1M concentration of KCl, KBr and KI are all found to fall on one
isotherm, as shown in Figure 2. These data indicate clearly that
anions have no influence.
    A number of points of clarification are now required to ex-
plain the specific differential effect of cations. The specific
adsorption of cations on the negatively charged cellose surface is
unlikely to be a controlling feature for adsorption, since the cations of the alkali metal halides are known to be only diffusely
adsorbed in the swollen aqueous fiber phase. Evidence for diffuse
adsorption of cations in the electrical double layer present in
cellulose-aqueous dye + electrolyte solution interfaces has been
obtained from the extensive investigations on Donnan potentials

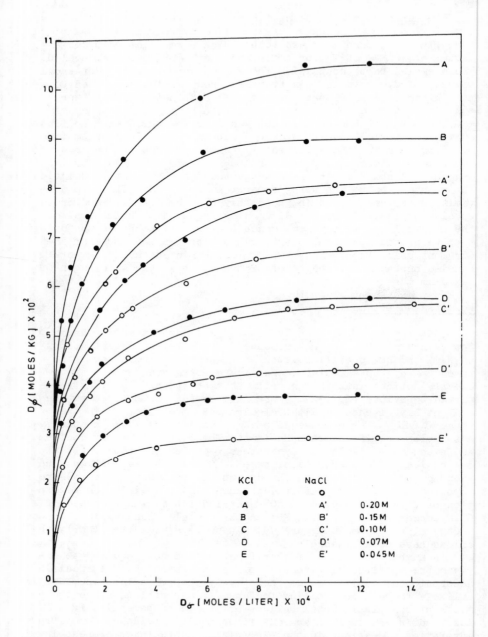

Figure 1. Adsorption isotherms for CSBFF on viscose at different concentrations of NaCl and KCl, at 50°C

and Donnan distribution of electrolyte ions in such systems using
different alkali halides ([4],[5],[6],[7],[8],[9],[10]), studies on negative
adsorption of chloride ions on cellulose fibre surfaces ([17]),
electrokinetic potentials at cellulose-aqueous KCl and NaCl so-
ltion systems ([19]) and a correlation between theory and experiment for the adsorption of direct dyes on cellulose ([1],[2],[3],[11],[20],
[21],[22]). Hence specific adsorption of the cations to different extents cannot explain the differences in dye anion uptake. The observed differences in the presence of NaCl and KCl cannot also be
attributed to differences in solubility of this dye since it has
been shown ([3]) that there are no significant changes in solubility
of the dye at a given electrolyte concentration.

However, a satisfactory explanation for the influence of the
type of electrolyte cation on dye adsorption by cellulose at
equivalent concentrations of NaCl and KCl can be obtained from a
consideration of the interactions with water of the different
species involved in the adsorption process. The cellulose surface
itself initially adsorbs water forming a layer of strongly bound
water molecules. Existence of structured water in the immediate
vicinity of the cellulose surface due to interactions between the
polar groups of the solid (viscose) and water is evident from the
work of Ramiah and Goring ([23]) and Drost-Hansen ([24]). Adsorption
of any other species will be hindered by the presence of this
vicinal water. In the solution phase, dissolved ions carry around
each of them a number of rigidly bound water molecules. These
hydrated ions by virtue of their strong electrostatic field tend
to disrupt the structure of water and increase the orientational
disorder of water surrounding their hydrated envelope. Frank and
Evans ([25]) called this effect the water structure breaking effect.
This effect increases with increasing ionic radius in the case of
cations ([26]), ([27]). When such hydrated ions are diffusely adsorbed in the electrical double layer adjacent to cellulose, they
can break the water-cellulose surface bonds or in other words, can
disrupt the adsorbed layer of water. In a similar manner these
cations can also break the ordered structure of water ("icebergs")
around non-polar parts of the dye anions ([28],[29],[30],[31],[32],[33]).
Such disturbances of the adsorbed water molecules on the cellulose
surface and the breakdown of "icebergs" around the dye molecule
will increase with increasing ionic strength. In effect, the dye
anions will approach the cellulose surface very closely thereby
enabling an increase in the intermolecular forces between them.
The greater water structure breaking effect of K^+ as compared to
Na^+ thus leads to the enhanced dye uptake in the presence of K^+
as shown in Figure 1.

Thermodynamics of Adsorption of CSBFF on Viscose. Isotherms,
after correcting for a Donnan equilibrum according to methods referred to earlier ([2],[3]), were found to give good linear reciprocal Langmuir plots. From the slopes and intercepts of these
plots, saturation values as well as standard thermodynamic param-

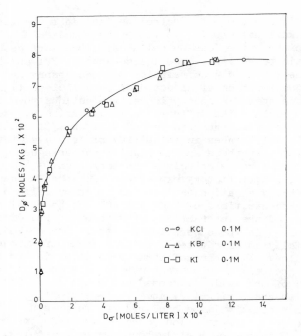

Figure 2. Adsorption isotherms for CSBFF on viscose in the presence of 0.1M concentrations of KCl, KBr, and KI at 50°C

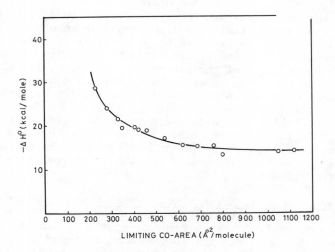

Figure 3. Variation of $-\Delta H°$ with limiting coarea of the adsorbed dye molecule on viscose

meters were calculated in the usual manner. Figure 3 which shows the variation of enthalpy changes $-\Delta H^O$ with limiting coarea per adsorbed dye molecule was constructed using the present as well as previous data (3). These results indicate that $-\Delta H^O$ is independent of the limiting coarea of the adsorbed dye molecule until about $6nm^2$. A gradual change is now seen to occur in the range of coareas between 6-3.5 nm^2 indicating increasing lateral dye-dye interactions. At about a coarea of 3.66 nm^2 for the adsorbed dye molecule, its orientation will correspond to a molecule lying flat on the surface with the plane of the benzene rings parallel to the surface. This limiting coarea for flat orientation was calculated from a Courtauld's atomic model of the dye. The further sharp rise in $-\Delta H^O$ shown in Figure 3 in the region from 3.5-2.0 nm^2 indicates increased stronger lateral dye-dye interactions. The changes in limiting coarea of the dye molecule from 3.5-2.0 nm^2 can be attributed to a further change in orientation of the adsorbed dye to one in which the plane of the benzene rings lies perpendicular to the surface. This latter orientation corresponds to a projected coarea of 2 nm^2 as determined from a Courtald's atomic model of the dye.

Adsorption of CSBFF on TiO$_2$. The adsorption isotherms of CSBFF on TiO$_2$ in the presence of NaCl and KCl at 35^OC and at 35° and 45^OC in the presence of KCl are given in Figures 4 and 5 respectively. It is interesting to note that the adsorption behaviour exhibited by cellulose is reflected strongly by TiO$_2$ also. In the absence of any electrolyte, TiO$_2$ does not adsorb CSBFF just as in the case of viscose. By including some NaCl in the dyebath slight adsorption takes place. However, the amount of dye uptake even at the high concentration of 0.2M NaCl is very small unlike in the case of viscose, where adsorption is considerable even at 0.045M NaCl. In contrast to this slight adsorption when NaCl is present, considerable dye adsorption takes place even in the presence of 0.1M KCl. Just as in the case of viscose, electrolyte anions Cl$^-$, Br$^-$ or I$^-$ were found to have no effect on dye adsorption.

The dye adsorption isotherm at 0.2M KCl is found to obey the Langmuir type equation and the adsorption process is markedly exothermic as can be noticed from the isotherms at 35° and 45^OC shown in Figure 5. From the reciprocal Langmuir plot for the isotherm at 35^OC in the presence of 0.2M KCl the coarea of the adsorbed dye molecule was calculated to be about 3.5nm^2. This corresponds to a flat orientation of adsorbed dye molecules on the TiO$_2$ surface indicating saturation adsorption. The same saturation on the viscose fiber surface is reached at a KCl concentration of 0.1M. These results show that TiO$_2$ which is a hydrophilic rigid surface, exhibits as in the case of viscose, the same characteristic behaviour towards dye adsorption in the presence of electrolytes. Hence the model suggested for explaining the dye adsorption behaviour on viscose in the presence of different

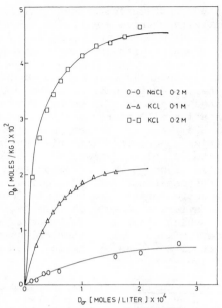

Figure 4.   Adsorption isotherms for CSBFF
on TiO_2 in the presence of NaCl and KCl at
35°C

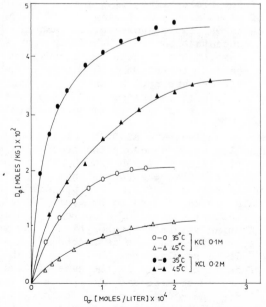

Figure 5.   Adsorption isotherms for CSBFF on
TiO_2 in the presence of 0.1M and 0.2M KCl at 35
and 45°C

electrolytes can be extended to TiO_2 also. The differences in the
magnitude of the cation effects on the adsorption of dye on. TiO_2
as compared to viscose can now be explained as follows:

TiO_2 has a strong affinity for water. From the heat of immersion studies carried out by Zettlemoyer et al. (34) the interaction of water with TiO_2 was found to have an energy of the order
of about 8-9 kcal/mole. This is much more than the energy of interaction of water with cellulosic material which falls within the
range 4.9-6 kcal/mole (14,35). Because of the stronger $TiO_2 - H_2O$
bonding, stronger forces are required to disrupt the structure
around TiO_2 as compared to viscose. In view of the observation
that the adsorption of the dye by TiO_2 is only of small magnitude
in the presence of even 0.2M NaCl it is surmised that the water
structure breaking effect of Na^+ ions is somewhat deficient to en-
courage dye adsorption. However, K^+ ions have a strong influence
and their effect is reflected in promoting saturation adsorption
on TiO_2. Evidence in support of this argument is obtained from
studies by Berube and DeBruyn (36) on the capacity of the electrical double layer in a TiO_2 - aqueous electrolyte solution system,
using KCl and NaCl as the electrolytes. These authors have
reached the same conclusions, namely, Na^+ will have little influence on water structure surrounding the TiO_2 surface whereas
K^+ ions disrupt it.

   Adsorption of CSBFF on SiO_2. Dye adsorption experiments were
carried out at electrolyte concentrations in the range 0 to 3.5 M
NaCl and KCl respectively. These experiments indicate that no
adsorption of the dye take place at any electrolyte concentration.
This is a significant observation different from the observations
made in the case of viscose or TiO_2. Silica is known to bind
water more strongly with an energy of the order of 9 to 20 kcal/
mole as calculated from heat of immersion studies (34,37). These
binding energy values are of much greater magnitude than for the
binding of water with cellulose and TiO_2. From a NMR study of
water on silica, Pickett and Rogers (38) observed very high surface coverage. The polarization and orientation of the first few
layers of water in turn causes further layers to be built up with
a lattice like order. Above 50 layers the ordered structure
breaks down leading to a more mobile gel-like structure. In view
of this very strong interaction of water with silica substrates
and the presence of a thick envelope of structured water layers
surrounding silica, the inability of the anionic dye to adsorb is
easily understandable. Evidently, the electrolyte cation cannot
influence this strongly bound water envelope surrounding silica
even at the highest concentration studied and hence cannot play
its usual role in favouring dye adsorption.

   Adsorption of CSBFF on Graphon. It can be noticed from
Figure 6 that the saturation value of dye uptake is the same
both in the absence and presence of electrolyte. Furthermore, the

type of electrolyte cation K⁺ or Na⁺ has no influence on the
saturation value or on the extent of equilibrium dye adsorption
at any stage of the isotherm, since the isotherms in the case of
KCl and NaCl overlap each other. The influence of electrolyte
is noticed only in the initial stages of the adsorption isotherm
and is understandable from the point of view of reduction, by the
electrolyte cation, of the electrostatic repulsion due to adsorbed
dye anions for other dye anions approaching the adsorbent surface.
As in the systems discussed earlier, anions did not influence dye
adsorption.

Figure 7 shows adsorption isotherms of CSBFF on Graphon in
the temperature range 30° to 50°C and 0.13 M NaCl. Here it can be
observed that the saturation value is unaffected by temperature,
but during the earlier stages of the dye adsorption isotherms that
there is a slight increase in equilibrium adsorption with temperature i.e. the adsorption process is endothermic. The isotherms
are of the Langmuir type and similar to those reported in the
literature for anionic dye adsorption on Graphon (39,40). The
reciprocal Langmuir plots for these isotherms are shown in Figure
8 and the calculated saturation value in the temperature range 30°
- 50°C again corresponds to a closepacked monolayer with a limiting coarea of 3.67 nm² per adsorbed dye molecule. This coarea as
discussed earlier, indicates an orientation of the adsorbed dye
molecule in a flat configuration with the benzene rings parallel
to the Graphon surface.

Thermodynamics of Adsorption of CSBFF on Graphon. From these
reciprocal Langmuir plots the standard thermodynamic parameters
shown in Table I have been calculated in the usual way. In this
Table, ΔH^o is seen to be a small positive value and the entropy
of adsorption is a large positive quantity. Similar positive
entropy values have been observed by Schneider et al. (41) for
adsorption of aliphatic acids and alcohols on polystyrene. These
results indicate that strong hydrophobic interactions are involved
in the adsorption of the anionic dye CSBFF on graphon. An important feature of hydrophobic bond formation discussed by Nemethy
and Scheraga (42), Kauzmann (43) and Schneider et al. (41), is
that the enthalpy of formation of the hydrophobic bond is positive
at low temperatures and becomes more negative passing through zero
as the temperature increases. This behaviour is logical since the
enthalpy of hydrophobic bond formation is a net effect of two
factors namely, the enthalpy of "iceberg" destruction and the
enthalpy of the bonding between the interacting species. The
first factor which is positive, predominates at low temperatures
and decreases with increasing temperature. The second factor is
negative and remains more or less unaffected in the narrow temperature ranges usually studied. Experimental evidence for these
enthalpy changes with increasing temperature has been provided by
Schneider et al. (41), who observed a decrease in adsorption at
higher temperatures in the case of polystyrene - hydrocarbon

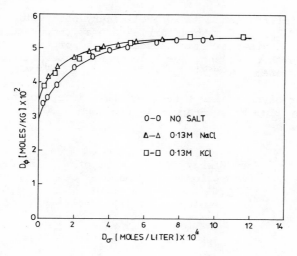

Figure 6. Adsorption isotherms for CSBFF on graphon in the absence and presence of electrolyte at 30°C

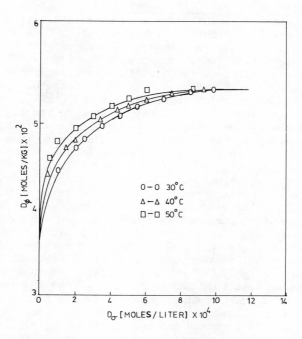

Figure 7. Adsorption isotherms for CSBFF on graphon at different temperatures in the presence of 0.13M NaCl

hydrophobic interactions. Some preliminary observations on the adsorption of dye on Graphon, carried out at temperatures higher than 50°C, indicate a similar decrease in adsorption. These results give further evidence for hydrophobic interactions in the Graphon-dye system.

Table I. Thermodynamic Parameters for the Adsorption of Chlorazol Sky Blue FF on Graphon in Presence of 0.13M NaCl.

Temperature (°C)	Free Energy of Adsorption ΔG^o (kcal/mole)	Heat of Adsorption ΔH^o (kcal/mole)	Entropy of Adsorption ΔS^o (e.u.)
30	−6.26		
40	−6.62	+4.65	+36
50	−6.98		

Adsorption of CSBFF on Activated Charcoal. Adsorption of CSBFF on activated charcoal presents a very interesting study. Figure 9 illustrates adsorption isotherms at 30°C on activated charcoal in the presence and absence of electrolytes in the dyebath. Two interesting features can be noticed from this figure. In the absence of electrolytes, activated charcoal shows considerable adsorption of CSBFF just as in the case of Graphon and in contrast to the behaviour observed for hydrophilic surfaces like viscose and TiO_2. The presence of electrolyte however, enhances the tendency for activated charcoal to adsorb the dye to a considerable extent. This behaviour is similar to that observed in the case of hydrophilic substrates but in contrast to that of Graphon. Thus activated charcoal exhibits a dual behaviour with respect to dye adsorption. In the absence of electrolytes, the Langmuir type adsorption tends to a limiting value of 3.5×10^{-2} moles/kg. Making the assumption that the adsorbed dye molecule occupies a limiting coarea of 3.67 nm^2 per molecule as in the case of Graphon, dye adsorption will correspond to a coverage of 80 m^2/g. This area is only 10% of the total surface and perhaps represents the totally hydrophobic surface of charcoal which is available for adsorption of large molecules. It is also reasonable to consider that this interaction is of the hydrophobic type as in the case of Graphon since a few experiments (not reported here), indicate a slight endothermic adsorption process. The dye uptake was found to increase enormously on adding electrolytes to the dyebath as shown in Figure 9. The presence of negatively

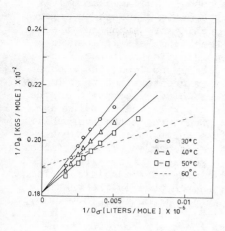

Figure 8. Reciprocal Langmuir plots. $1/D_\phi$ vs. $1/D_\sigma$ for the adsorption of CSBFF on graphon at different temperatures. NaCl 0.13M. D_ϕ and D_σ refer to the amount of dye adsorbed and the equilibrium dyebath concentration respectively.

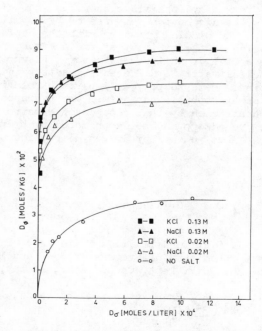

Figure 9.   Adsorption isotherms for CSBFF on activated charcoal in the presence and absence of electrolyte at 30°C

charged acidic sites on activated carbons including charcoal is well known $(\underline{39},\underline{44},\underline{45})$. The electrolyte cations screen the electrostatic repulsion between the approaching dye anion and the negatively charged acidic sites and hence enable a greater adsorption of dye anions. The differential effects of the electrolyte cations K^+ and Na^+ are also clearly noticed, the adsorption in the presence of K^+ being always greater than that in the presence of Na^+. The observed increase in dye adsorption in the presence of K^+ as compared to Na^+ can be explained as being due to a greater disruption of the water structure on the hydrophilic parts of the charcoal surface by K^+ ions and parallels the behaviour for dye adsorption on hydrophilic viscose and TiO_2 surfaces. The hydrophilic character of charcoal surfaces arises from the ordering of water molecules in the vicinity of the acidic surface groups $(\underline{46},\underline{47},\underline{48})$. For the charcoal sample used in the present study, the acidic groups approximately correspond to 0.8 mequiv/g

The present investigations using the dye adsorption technique, bring to light not only the variations in the interfacial behaviour of different hydrophilic substrates, but also the differences between the adsorption features of hydrophilic and hydrophobic surfaces. The binding of water with the surface and its further structuring in the vicinity of the surface are characteristic features of hydrophilic substrates. Disruption of this structure of vicinal water to different extents by different electrolyte cations seems to play an important role in regulating dye adsorption on hydrophilic surfaces. The hydrophobic surface is practically unaffected by the presence of electrolytes in the aqueous medium, dye adsorption being promoted by hydrophobic interactions that involve entropy driven disruption of "iceberg" water.

Literature Cited

1.   Sivaraja Iyer, S.R., Srinivasan, G., Baddi, N.T., and
     Ravikrishnan, M.R., Text. Res. J., (1964), $\underline{34}$, 807.

2.   Sivaraja Iyer, S.R., and Baddi, N.T., in Proc. Symp. "Contributions to Chemistry of Synthetics Dyes and Mechanism of Dyeing", p. 36, Univ. Dept. Chem. Tech., Bombay, India (1967)

3.   Sivaraja Iyer, S.R., Srinivasan, G. and Baddi, N.T., Text. Res. J., (1968), $\underline{38}$, 693.

4.   Neale, S.M., Trans. Faraday Soc., (1947), $\underline{43}$, 325.

5. Neale, S.M., and Standring, P.T., Proc. Roy. Soc. London, Ser. A, (1952), 213, 530.

6. Neale, S.M., and Saha, P.K., J. Soc. Dyers Colour., (1957), 73, 381.

7. Sivaraja Iyer, S.R., and Jayaram, R., J. Soc. Dyers Colour., (1970), 86, 398.

8. Usher, F.L., and Wahbi, A.K., J. Soc. Dyers Colour., (1942), 58, 221.

9. Neale, S.M., and Farrar, J., J. Colloid Sci., (1952), 7, 186.

10. Sivaraja Iyer, S.R., and Kalbag, V.N., Unpublished results.

11. Hanson, J., Neale, S.M. and Stringfellow, W.A., Trans. Faraday Soc., (1935), 31, 1738.

12. Vickerstaff, T., "The Physical Chemistry of Dyeing", Oliver and Boyd, London, 1954.

13. Sivaraja Iyer, S.R. in "The Chemistry of Synthetic Dyes," K. Venkataraman, Ed., Vol. VII, Chap. IV, Academic Press, New York, in press.

14. Stamm, A.J., "Wood and Cellulose Science", p. 248, The Reinhold Press Co., New York, (1964).

15. Robinson, C., and Mills, H.A.T., Proc. Roy. Soc. London, Ser. A, (1931), 131, 596.

16. Whistler, R.L., Ed., "Methods in Carbohydrate Chemistry", Vol. 3, p. 3 Academic Press, New York, 1963.

17. Nemade, B.I., Sivaraja Iyer, S.R. and Jayaram, R., Text. Res. J., (1970), 40, 1050.

18. Zettlemoyer, A.C., in "Hydrophobic Surfaces", F.M. Fowkes, Ed., p. 1, Academic Press, New York, 1969.

19. Sivaraja Iyer, S.R., and Jayaram, R., J. Soc. Dyers Colour., (1971), 87, 338.

20. Crank, J., J. Soc. Dyers Colour, (1947), 63, 293.

21. Peters, R.H. and Vickerstaff, T., Proc. Roy. Soc. London Ser. A, (1948), 192, 292.

22. Standing, H.A., and Warwicker, J.O., J. Text. Inst., (1949), 40, T175.

23. Ramiah, M.B. and Goring, D.A.I., J. Poly. Sci. Part C, 2, (1965), 27.

24. Drost-Hansen, W., Ind. Eng. Chem., (1969), 61(11),10

25. Frank, H.S., and Evans, M.W., J. Chem. Phys., (1945), 13, 507.

26. Frank, H.S., and Wen, W.Y., Disc. Faraday Soc., (1957), 24, 133.

27. Wicke, E., Angew. Chem. Internat. Edit., (1966), 5, 106.

28. Sivaraja Iyer, S.R., and Singh, G.S., Kolloid-Z.Z. Polym, (1970), 242, 1196.

29. Nemethy, G., Angew. Chem. Internat. Edit., (1967), 6, 195.

30. Mukerjee, P., and Ghosh, A.K., J. Phys. Chem., (1963), 67, 193.

31. Rohatgi, K.K., and Singhal, G.S., J. Phys. Chem., (1966), 70, 1695.

32. Katayama, A. Takagishi, T., Konishi, K., and Kuroki, N., Kolloid-Z.Z. Polym, (1965), 206, 162.

33. Sivaraja Iyer, S.R., and Singh, G.S., J. Soc. Dyers Colour, (1973), 89, 128.

34. Zettlemoyer, A.C., and Chessick, J.J., Advan. Catalysis, (1959), 11, 263.

35. Baddi, N.T., Cell. Chem. Tech., (1969), 3, 561.

36. Berube, Y.G., and DeBruyn, P.L., J. Colloid and Interface Sci., (1968), 28, 92.

37. Mackrides, A.C., and Hackerman, N., J. Phys. Chem., (1959), 63, 594.

38. Pickett, J.H., and Rogers, L.B., Anal. Chem., (1967), 39, 1892.

39. Graham, D., J. Phys. Chem., (1955), 59, 896.

40. Nandi, S.P. and Walker, Jr. P.L., Fuel, (1971), 50, 345.

41. Schneider, H., Kresheck, G.C., and Scheraga, H.A., J. Phys. Chem. (1965), 69, 1310.

42. Nemethy, G., and Scheraga, H.A., J. Phys. Chem., (1962), 66, 1773.

43. Kauzmann, W., Adv. Protein Chem., (1959), in 14, 1.

44. Coughlin, R.W., Ezra, F.S., and Tan, R.N., in "Hydrophobic Surfaces", F.M. Fowkes, Ed., p. 44. Academic Press, New York, 1969.

45. Boehm, H.P., Diehl, E., Heck., W., and Sappock, R., Angew. Chem. Internat. Edit., (1964), 3, 669.

46. Walker, Jr. P.L., and Janov, J., in "Hydrophobic Surfaces" F.M. Fowkes, Ed., p. 107, Academic Press, New York 1969.

47. Dubinin, M.M. in "Chemistry and Physics of Carbon", P.L. Walker, Jr., Ed., Vol. 2, p. 51, Marcel - Dekker, New York 1965.

48. Mattson, J.S., and Mark Jr., H.B., "Activated Carbon - Surface Chemistry and Adsorption from Solution", Marcel-Dekker, New York, 1971.

7

Adsorption of Polystyrene on Graphon from Toluene

VICTOR K. DUNN* and ROBERT D. VOLD

Department of Chemistry, University of Southern California, Los Angeles, Calif. 90007

Introduction

Recently many investigators have been concerned with the stability of non-aqueous suspensions, and a number of alternative
hypotheses have been proposed to account for the observed facts
(1,2,3). All involve consideration of the overlap region of the
adsorbed stabilizer surrounding the suspended particle, and predict an increase in stability with increasing molecular weight
of an adsorbed macromolecule. In connection with a project to
determine quantitatively the effect of the molecular weight of
adsorbed polystyrene on the stability of Graphon suspensions in
toluene, it therefore became necessary to obtain data on the
extent of adsorption and the thickness of the adsorbed layer
under the same identical conditions under which subsequent determinations of the rate of flocculation of the suspension could be
carried out. Such data are also useful for comparison with the
predictions of computer simulations of adsorption of macromolecules (4).

Determination of adsorption on Graphon in suspension is
greatly complicated by the fact that the material is present as
aggregates rather than primary particles. Hence the amount of
polystyrene adsorbed will depend on the degree of dispersion of
the Graphon, which in turn is a function of the level of mechanical agitation, and also dependent on whether the agitation
takes place in the presence or absence of the polystyrene. Since
it was desired to compare stabilities at saturation adsorption of
a number of polystyrene samples of different molecular weights,
it was first necessary to determine adsorption isotherms for each
of the polystyrenes so as to insure use of an initial concentration which would result in saturation adsorption. Since flocculation measurements were to be carried out in Graphon suspensions it was also necessary to study the rate of adsorption and

---

*Present address: Xerox Corporation, Webster Research Center,
W130, Rochester, New York 14644.

determine the time required to reach at least approximately the
equilibrium value.

The results obtained yield interesting information concerning the probable conformation of the adsorbed molecules and the
dependence of the thickness of the adsorbed layer on the molecular weight of the polystyrene.

Materials and Methods

Materials. Graphitized carbon black (Spheron 6), prepared by
heating in the absence of air to 2700°C, or higher, was obtained
from the Cabot Corporation. Samples were extracted continuously
for 48 hours in a Soxhlet extractor with cyclohexane and toluene
successively. After this purification neither water nor toluene
equilibrated with the extracted Graphon showed any absorption
from 340 to 800 nm indicating that extractable impurities adsorbing in the visible region of the spectrum had been removed. The
surface area of the extracted Graphon was found to be 95.4 m^2/g
by low temperature adsorption of nitrogen by the B.E.T. method,
using 16.2Å^2 as the area per nitrogen molecule.

Toluene (Analytical Reagent, Mallinckrodt Co.) was distilled
through a 1-1/2 foot column (Allihn condenser) and collected
between 110.0 to 110.6°C. Since it had been shown (5) that non-
aqueous dispersions of powdered quartz were destabilized by
traces of water, it was necessary to work under strictly anhydrous conditions. Accordingly, about three feet of sodium wire
of 2 mm diameter was pressed into a gallon of freshly distilled
toluene and left for 24 hours, and the toluene thereafter stored
over sodium wire. Glassware used in the experiments was dried
at 160°C and used immediately after drying. Precautions to maintain absolute dryness were necessary, not only because of the
destabilization possible because of capillarity effects between
water films on the particles, but also because any such film
could make possible ionization of surface impurities in the
Graphon, such as carboxyl groups, and result in electrostatic
charges on the Graphon particles.

Polystyrene was obtained from the Pressure Chemical Company.
Five samples of nominal molecular weights* 20,400, 110,000,
200,000, 498,000 and 1,800,000 were used. For the first three
samples the ratio of weight average to number average molecular
weight is given as less than 1.06, and less than 1.20 for the
last two samples. Several of the preliminary experiments were
carried out with Dow Resin P-65, a polystyrene of broad molecular weight distribution and nominal molecular weight 2.3×10^5 as
determined from viscosity measurements. Samples were used directly from the bottle since there was no detectable change in
weight (within 0.00002 g) after drying two hours at 112°C and
10^{-3} torr.

*Weight average molecular weights of these samples are reported
to be 20,400; 111,000; 200,000; 507,000; and 1.9×10^6.

Determination of Adsorption Isotherms. The amount of adsorption
was determined from the change in concentration of polystyrene
in the Graphon suspensions in toluene prepared by ultrasonic
agitation after equilibration in the presence of varying concentrations of polystyrene. Since the amount adsorbed is dependent
on the degree of dispersion of the Graphon, the time allowed for
attainment of equilibrium, and the amount of agitation during
this period, extensive preliminary experiments were necessary to
establish a satisfactory procedure which could be used in subsequent experiments on the stability of the suspensions.

The procedure adopted for the final determinations was as
follows. The Graphon sample was outgassed for 24 hours at $160^{o}C$,
at 10^{-3} torr. A 0.5000 g sample was then weighed into a 125 ml
Erlenmeyer flask, followed by 25.0 ml of toluene. Each flask
was individually irradiated for 20 minutes in a 100 watt, 15 KHz
Delta Sonic ultrasonic tank with the selector switch at high
intensity. A 25.0 ml aliquot of a solution of known concentration of polystyrene in toluene was added to each flask immediately following irradiation, and the flasks then rotated three minutes at 56 r.p.m. at an angle of 30° to the vertical by a Borg
Equipment Division model 1007-4N rotor. They were then left undisturbed in a constant temperature room $(24^{o} \pm 1^{o}C)$ for 14 days.
The supernatant liquid was then decanted, and about 30 ml centrifuged for ten minutes at 4800 r.p.m. in an International Clinical
Centrifuge, model CL, to remove residual Graphon Particles. The
centrifuged liquid was clear and colorless and had zero absorbance at 710 nm using toluene as a reference liquid. The concentration of polystyrene after adsorption was determined by
placing 25.0 ml of the centrifuged liquid recovered from the dispersion in weighed "boats" made from aluminum foil, and drying
to constant weight $(\pm 0.05$ mg) at $112^{o}C$. The "boats", made by
folding the foil, were about 1" high and 3" by 3" square at the
base. About four to five hours were required to reach constant
weight.

During the equilibration the flasks were closed with two
layers of Saran wrap and a layer of aluminum foil held in place
by a rubber band around the neck. Since similarly closed flasks
containing initially 50 ml of toluene lost an average of only
0.2523 g weight in fourteen days, it is apparent that any change
in concentration of the polystyrene due to evaporation of toluene
is negligible under these conditions.

The results of the preliminary experiments carried out to
establish this final procedure are of value in themselves for
the insight they afford into the effect on adsorption of macromolecules on dispersed particles of the time allowed for equilibration, order of addition of reagents, degree of agitation of
the suspensions, and degree of dispersion of the Graphon. Figure
1 shows the effect of the degree of dispersion of Graphon at 23
$\pm 1^{o}C$ in a system prepared by irradiation of 0.5000 g of Graphon
in 25.0 ml of toluene for varying lengths of time, followed by

addition of 25.0 ml of a 4.00 mg/ml toluene solution of Dow
Resin P-65. The suspension was swirled gently for three minutes
and left to equilibrate undisturbed for seven days. Each point
shown in Figure 1 is the average from three separate experiments.

Up to 15 minutes the amount adsorbed increases significantly
with increasing time of irradiation, as might be expected from
the presumed decrease in average aggregate size. Further irradiation beyond 15 minutes had little effect, the amount adsorbed
remaining fairly constant at 45.3 to 46.5 mg polystyrene per
gram of carbon. It can therefore be assumed that after 15 minutes no additional subdivision of aggregates occurs with further
irradiation, analogous to earlier results (6) showing attainment
of a reversible equilibrum particle size distribution of carbon
and ferric oxide in aqueous suspensions dependent on the intensity of agitation. Accordingly, it was concluded that a 20
minute period of irradiation of the suspensions was sufficient
to insure reaching the limiting value of the adsorption.

The full circle represents the amount of adsorption found
when the solutions were swirled three minutes daily during the
equilibration period rather than being left undisturbed. The
increase in adsorption shows that even under these mild conditions some additional deflocculation occurred as a result of the
agitation in the presence of the polystyrene, as was also found
by other workers (7) who reported increases in adsorption of
polymer with agitation.

It is significant that even with no ultrasonic irradiation
of the suspension 39.4 mg of polystyrene were adsorbed per gram
of Graphon. This is about 85% of the limiting adsorption reached
after 20 minutes of irradiation, and demonstrates that polystyrene can reach most of surface area of the Graphon in suspension even though most of the particles are present in aggregates.

The precision of the measurements improved with longer periods of irradiation up to 20 minutes as shown in Table I. This
Table gives the relative error (defined in terms of the average
deviation from the average value) in the weight of polystyrene
adsorbed per gram of Graphon at each time of irradiation for a
suspension containing initially 2 mg of polystyrene (Dow Resin
P-65) per ml of toluene. After irradiation the samples were
left seven days without agitation before determination of the
amount adsorbed, except for one set which were swirled for three
minutes daily. Apparently it is more difficult to reproduce a
given degree of dispersion at short periods of irradiation than
at longer. The reproducibility was much poorer for samples which
were agitated after addition of the polystyrene solution to the
Graphon suspension, presumably because the polystyrene could coat
and stabilize any fresh surface resulting from the agitation,
thus increasing the surface area of Graphon available for adsorption.

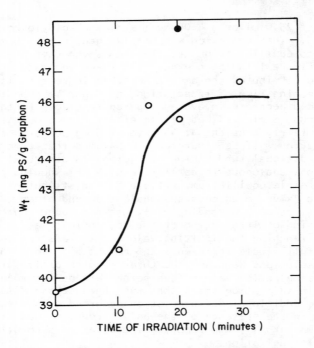

Figure 1. Adsorption of polystyrene (2 mg/ml, M.W. = 2.3 × 10⁵) onto Graphon from toluene as a function of time of ultrasonic irradiation. Time of equilibrium, 7 days. ○, no agitation; ●, 3 minutes of agitation daily.

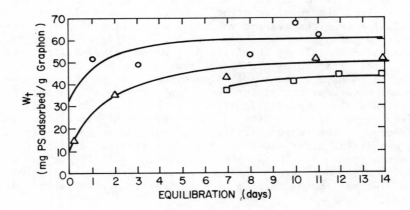

Figure 2. Adsorption of polystyrene (1 mg/ml) onto Graphon from toluene, ○, M.W. = 1,800,000; □, M.W. = 498,000; △, M.W. = 250,000, broad distribution

Table I. Precision as a Function of Irradiation Time

Irradiation Time (minutes)	0	10	15	20	30	20*
Average % deviation	1.6	4.5	3.0	1.4	2.2	8.0

   *These samples were swirled 3 minutes daily after irradiation

Experiments such as these led to the adoption of a standard procedure involving irradiation of the suspension for 20 minutes, and no further agitation after mixing the polystyrene solution with the Graphon suspension. It should be noted that irradiation must be carried out prior to addition of polystyrene. When Graphon suspensions containing polystyrene were irradiated for half a minute it was found that those containing polystyrene of higher molecular weight had a much higher absorbance, indicative of a smaller average aggregate size. Consequently no direct determination of the dependence of the amount adsorbed on the molecular weight of the polystyrene would be possible on suspensions so prepared because of the differences in their initial state.

Using the standard method of preparation of suspensions and determination of amount adsorbed already described, several experiments were performed to determine the amount of adsorption as a function of time of equilibration and of the molecular weight of the polystyrene. These data showed how much time had to be allowed to reach the saturation value of the adsorption. On the basis of all these preliminary experiments it was then possible to specify for the polystyrene samples of different molecular weights an initial concentration such that the equilibrium concentration remaining in the solution would be sufficient to maintain monolayer coverage of the Graphon.

Results and Discussion

Data on the rate of adsorption for polystyrene of high, low and intermediate molecular weight are shown in Figure 2. Few measurements were made at short periods of equilibration since the primary purpose of these experiments was to establish the length of time necessary to reach apparent adsorption equilibrium. Where studied, adsorption was relatively rapid during the first two days, amounting to about 70% of the quantity adsorbed after fourteen days. The limiting value of the amount adsorbed appears to have been reached after ten days equilibration with all three polystyrenes.

The adsorption isotherms of polystyrenes of differing molecular weights as determined by our standard procedure are shown in Figure 3. The amount adsorbed at first increases rapidly with concentration, and then levels to a plateau at an equilibrium concentration of polystyrene of 1 mg/ml. The shape of these isotherms is suggestive of monolayer adsorption, and resembles that found by Schick and Harvey (8). They studied the adsorption of a polystyrene of M_w 292,000 onto Graphon from several solvents,

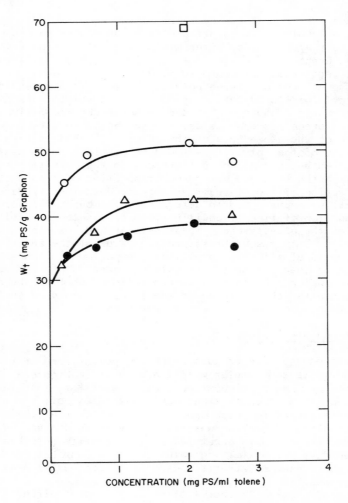

*Figure 3.   Adsorption of polystyrene from toluene onto Gra-
phon. ●, M.W. = 110,000; △, M.W. = 200,000; ○, M.W. =
498,000; □, M.W. = 1,800,000.*

including toluene, and also found that a limiting adsorption
characteristic of monolayer adsorption was reached. The present
work shows that the maximum amount of polystyrene adsorbed on
Graphon increases from 39 to 69 mg/g as the molecular weight increases from 110,000 to 1,800,000.

In Figure 4 the saturation adsorption is plotted against the
square root of the molecular weight of the adsorbed polystyrene.
The amount adsorbed varies linearly with the square root of the
molecular weight according to the equation,

$$W_t = 3.0 \times 10^{-2} \sqrt{M} + 28 \qquad (1)$$

where W_t is mg of polystyrene adsorbed per gram of Graphon and
M is the molecular weight of the polystyrene. The line extrap-
olates to a non-zero intercept, a not uncommon result in the
adsorption of polymers (9,10).

The fact that the adsorption data can be represented by
Equation (1) leads to some interesting conclusions concerning
the conformation of the adsorbed polystyrene. As extreme
limiting cases the polymer molecule could be held completely flat
on the surface, or it could be present as essentially rigid rods
in close packing oriented perpendicularly to the surface. If the
molecules are flat the weight adsorbed at saturation should be
very nearly independent of the molecular weight, whereas in a
perpendicular orientation it would be directly proportional to
the molecular weight. Since the data given in Figure 4 show that
neither of these predictions is correct, it can be concluded
that the conformation of the adsorbed molecule must be somewhere
in between the two extreme situations. Another possibility would
be for the molecule to be anchored terminally but to have a random coil configuration in solution. This would require that the
amount adsorbed be directly proportional to the square root of
the molecular weight, which is contradicted by the non-zero intercept indicated by Equation (1).

A model closely resembling those illustrated by Eirich et al
(10), which is in accord with the empirical equation representing
the data, is shown in Figure 5. The adsorbed layer may be divided into two regions. In region (1) the monomer segments are
anchored flat directly on the surface, the weight of segments
adsorbed in this fashion being indicated as W_0. In region (2)
the segments are present as loops or tails in the solution not
in actual contact with the surface, the weight so adsorbed being
indicated as W. Accordingly, if W_t is the total weight of polymer adsorbed, it follows that

$$W_t = W + W_0 \qquad (2)$$

By comparison with Equation (1) is seen that $W_0 = 28$ mg/g and
$W = 3.0 \times 10^{-2}$ M. Therefore, the fraction of adsorbed weight
independent of the molecular weight of the polystyrene is

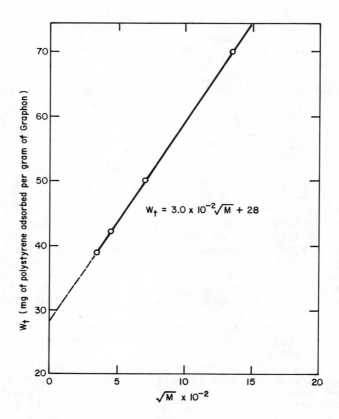

$$W_t = 3.0 \times 10^{-2}\sqrt{M} + 28$$

Figure 5. Hypothetical conformation of adsorbed polystyrene. ○, monomer segment.

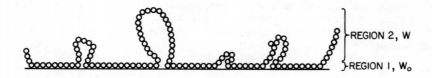

Figure 4. The square root of the molecular weight of polystyrene vs. the maximum amount of polystyrene adsorbed from toluene onto 1 g of Graphon

probably due to a constant amount of closely packed polymer segments that lie directly on the surface. The other fraction of
adsorbed weight consists of polymer segments in the form of loops
and tails extending into the solution. The number and size of
the loops and tails increase with increasing molecular weight.
        Further verification of the conformation of the adsorbed
molecules, and the thickness of the adsorbed layer, can be obtained from the dependence of the effective surface area of the
adsorbed polystyrene on its molecular weight. Since the surface
area of the Graphon is 95.4 m^2/g, adsorption of 28 mg of polystyrene per gram corresponds to an effective surface area of
58.8Å^2 per monomer unit. Since it was known (11) from adsorption
studies of tetradecylbenzene on Graphon from heptane that the
effective surface area of a benzene ring and one methylene group
flat on the surface in a saturated monolayer is 50Å^2, this indicates that the polystyrene segments here adsorbed flat on the
surface are present in a closely packed layer. That the weight
of polystyrene adsorbed as unanchored segments in region (2)
varies as the **square** root of the molecular weight suggests that
the number of loops and their extension into the solution, and
hence the thickness of the adsorbed film, increase with increasing molecular weight.

Table II. Fraction of Adsorbed Polystyrene Not Bound to the
          Surface

Mol. Wt.	Wt. of Polystyrene Adsorbed at Saturation mg/g of Carbon	Apparent Area per Monomer,Å^2	Wt.Fraction Not Directly Bound on Surface
*	28.0	58.8	0.00
110,000	39.0	42.2	0.28
200,000	42.5	38.8	0.34
498,000	50.0	32.9	0.44
1,800,000	69.8	23.6	0.60

* Extrapolated adsorption of 28 mg/g at zero molecular weight from
Figure 4.

        This conclusion is further supported by the calculations
shown in Table II. At saturation adsorption the apparent area
available per adsorbed monomer unit decreased from 42.2Å^2 to
23.6Å^2 as the molecular weight increased from 110,000 to
1,800,000. Since the actual area per adsorbed segment would
remain constant in a saturated monolayer, this result shows that
an increasing number of segments no longer lie flat on the surface as the polystyrene molecule becomes bigger. According to
Equations (1) and (2) and the model of Figure 5, 28 mg of polystyrene per gram of Graphon are flat on the surface with the
remainder present as unbound segments. Subtraction of 28 mg from
the saturation adsorption in mg/g indicated by the isotherms of

Figure 3 gives the weight of adsorbed polystyrene segments not flat on the surface, and so permits calculation of the weight fraction of unbound segments as shown in Table 2. The increase in value with increasing molecular weight from 28% to 110,000 to 60% at 1,800,000 is further indication of the increasing thickness of the adsorbed layer with increasing molecular weight.

Summary

The results reported show that fourteen days are sufficient to reach adsorption equilibrium between Graphon and a solution of polystyrene in toluene, and that an equilibrium concentration of 1 mg/ml is sufficient to attain saturation adsorption. The dependence of the saturation adsorption on the molecular weight of the polystyrene indicates that at low molecular weights the polystyrene is mostly present flat on the surface, with loops extending progressively further out into the solution as the molecular weight increases.

Literature Cited

1. Clayfield, E.J., and Lumb, E.C., J. Colloid Interface Sci., (1966), 22, 269, 285.
2. Ottewill, R.H., Kolloid-Z.Z. Polym. (1967), 227, 108.
3. Bagchi, P. and Vold, R.D., J. Colloid Interface Sci. (1970), 33, 405.
4. Clayfield, E.J. and Lumb, E.C., J. Colloid Interface Sci., (1974), 47, 6, 16.
5. Kruyt, H.R., and van Selms, F.G., Rec. Trav. Chim. (1943), 62, 407, 415.
6. Reich, I., and Vold, R.D., J. Phys. Chem. (1959), 63, 1497.
7. LaMer, V.K. and Healy, T.W., Rev. Pure Applied Chem. (Australia) (1963), 13, 112.
8. Schick, M.J. and Harvey, E.N., in "Interaction of Liquids at Solid Substrates," Adv. Chem. Ser., No. 87, p. 63, American Chemical Society, Washington, D.C., 1968.
9. Fontana, B.J. in "The Chemistry of Biosurfaces", M. Hair, Ed., Vol. I, p. 83, Marcel Dekker, Inc., New York, 1971.
10. Eirich, F.R., Bulas, R., Rothstein, E., and Rowland, F., in Chemistry and Physics of Interfaces, D. E. Gushee, Ed., p.109, American Chemical Society, Washington, D.C., 1965.
11. Parfitt, G.D., and Willis, E., J. Phys. Chem. (1964) 68, 1780.

Adsorption of Ionic Surfactants to Porous Glass: The Exclusion of Micelles and Other Solutes from Adsorbed Layers and the Problem of Adsorption Maxima

PASUPATI MUKERJEE and AROONSRI ANAVIL

School of Pharmacy, University of Wisconsin, Madison, Wis. 53706

Introduction

Porous glass, containing a rigid, interconnecting network of pores (1), is used extensively for chromatographic separations of polymers (2), proteins (3), polysaccharides (4), and viruses (1,5). Because of the presence of pores, these glasses have high surface areas per unit weight (6,7). The adsorption of surfactants to glass (8) is of interest for a variety of reasons. The present work is concerned with the adsorption of cationic and anionic surfactants to porous glass with special attention to concentrations above the critical micellization concentration (c.m.c.). The work is expected to provide background information for chromatographic studies on micellar systems. Ionic micelles are surrounded by electrical double layers, the effective extent of which depends mainly upon the surface charge density and the ionic strength. The dimensions of the electrical double layers around charged surfaces are also similarly determined. It was hoped that a study of the adsorption to porous glass containing pores that are rigid and large compared to molecular dimensions, the interactions of micelles with adsorbed layers could be critically studied.

The interpretation of adsorption isotherms of ionic surfactants to solid or liquid substrates as also "binding" isotherms to macromolecules involves a variety of factors. These include adsorbate-adsorbent interactions as also adsorbate-adsorbate interactions in the adsorbed layers. The range of an interaction may vary with its type. The nonpolar moiety of the adsorbed molecule is expected to show mostly short-range interactions. The charge interactions, however, can be of short-range as also long range. For short-range interactions, the existence of narrow pores and/or surface nonuniformities of the order of molecular dimensions can produce important curvature effects even in an otherwise homogeneous solid. The characteristic distance of the long-range electrical double layer (approximately $0.3/\sqrt{C}$ nm where C is the bulk concentration of a 1:1 electrolyte). Since this Debye thickness can be much greater than molecular dimensions, the presence

of fairly wide pores and/or relatively gross nonuniformities at
the surface of nonporous solids can produce curvature effects for
electrical double layers, and thus affect adsorption.

A second set of problems we would like to draw attention to in
this study arises from the interactions of bulk solute species
with the adsorbed layer. These problems are particularly important when the amount adsorbed is calculated from the change in
solute concentration of bulk solutions far from adsorbed layers.
For example, it is well known that co-ions are excluded from the
electrical double layer ($\underline{9}$, $\underline{10}$). Its importance in the calculation of the true amount adsorbed, i.e., the amount at the surface
where the adsorbed molecules are within the field of short-range
interactions with the surface, has been indicated by Van Dolsen
and Vold ($\underline{11}$). In a typical case of the adsorption of an ionic
surfactant to an initially uncharged surface, the concentration of
the surfactant ion in the bulk solution far from the surface is
increased because of its expulsion from the electrical double
layer adjacent to the adsorbed layer. This results in an under-
estimate of the true amount adsorbed at the surface. The classical Donnan type corrections for colloidal systems are based on
this principle but use idealizations of the system which are not
acceptable for highly charged surfaces ($\underline{12}$).

If the adsorption experiment is so conducted that a finely
divided phase is allowed to sediment or is centrifuged out before
concentration measurements are performed on the continuous phase,
the problem becomes more complex because the interparticle separations between the particles of the separated phase may be controlled in part by geometrical factors such as asperities, long-
range van der Waals' forces, gravitational forces for coarse particles, as also the repulsive interactions of the electrical double
layers. To calculate the true amount adsorbed at the surface, it
is necessary to take into account the negative adsorption of co-
ions from an interacting double layer system under the influence
of the other forces mentioned. If the adsorbent phase is separated too rapidly, nonequilibrium situations could also arise.

In micelle-forming systems, further complications arise from
the interaction of ionic micelles with adsorbed layers. Ionic
micelles with their electrical double layers are unlikely to show
positive adsorption to uncharged or similarly charged surfaces.
On the other hand, repulsive interactions between similarly
charged surfaces and the highly charged surfactant micelles are
expected from double layer theory and may be pronounced in view of
the pronounced repulsive interactions exhibited by the micelles
themselves ($\underline{13}$). As in the case of small co-ions, such interactions should result in an excluded volume type interaction which
is expected to give higher concentrations of micelles far from the
adsorbed layers than close to it, thus resulting in an underesti-
mate of the true adsorption of the surfactant. When a finely divided adsorbent phase is separated from the solution phase, the
interactions of the ionic micelles with the particles of the

adsorbent phase mutually interacting with each other is expected
to produce a multiplicity of effects involving changes in the
interparticle separations at equilibrium, the negative adsorption
of both co-ions and micelles, as also the electrical interactions
in the adsorbed layers, and possible changes in the monomer-
micelle equilibrium.

One purpose of the present work was to investigate the nature
of the adsorption isotherms of micelle forming surfactants above
the c.m.c. in some suitably chosen systems designed, in particular, to examine the significance of micellar exclusion from similarly charged surfaces, and to what extent these exclusion effects
may be involved in the interpretation of the problem of adsorption
isotherms showing maxima above the c.m.c. (14,15). The adsorption
substrate used was porous glass (Bio-Glas) containing pores of
average diameters of about 16 nm and 35 nm. Because of the
presence of an extensive, interconnecting network of pores, these
glasses have high surface areas per unit weight of solid even for
coarse particles. Electron micrographs (16) indicate highly un-
even, pitted surfaces, so that a significant fraction of the surface may be external. Because of the rigidity of glass, and large
distances between external surfaces, all interactions between adsorption surfaces could be assumed to occur at constant geometry,
independent of the composition of the solution.

Experimental

Materials. Bio-Glas samples were obtained from Bio-Rad Laboratories. Two different samples of Bio-Glas 200, numbered I and II
(Table I) with a nominal exclusion limit of 20 nm and Bio-Glas
500, with a nominal exclusion limit of 50 nm were used. The glass
powders were of 50-100 mesh size, corresponding to coarse particles of roughly 150-300 μm in diameter. The particles were
irregular in shape, and settled rapidly from suspensions.

The manufacturer's figures showing the penetration of mercury
into a powder as a function of pressure indicated in all cases one
distribution of pore sizes from roughly 20-70 μm corresponding to
the external surface. A second pore size distribution centered
around roughly 16 nm for Bio-Glas 200 and 35 nm for Bio-Glas 500
corresponding to the internal surface. Table I indicates some
pore size parameters for the internal pores from the mercury penetration data. Adsorption measurements with cationic surfactants
indicated that different lot numbers had different surface areas
(Table I). Different bottles of the same lot number usually gave
concordant results in the adsorption experiments excepting in the
case of sodium tetradecyl sulfate for which different bottles of
Bio-Glas 200 (I) with the same lot number sometimes gave consistent differences in the adsorption, the maximum difference being
about 9×10^{-6} equiv./g, or about 15% of the maximum adsorption.
All the experimental data reported in a particular figure are based
on Bio-Glas from the same bottle, so that internal consistency of

Table 1. Characteristics of Porous Glass Samples Used

	Range of[a] pore diameters (nm)	Average[b] pore diameter (nm)	Internal[c] pore volume (ml/g)	Adsorption[d] at the c.m.c. (mole/g)	Surface[e] area (m^2/g)
Bio-Glas 200 (I)	14-25	18.5	0.67	3.9×10^{-4}	140
Bio-Glas 200 (II)	13-25	15	0.40	2.25×10^{-4}	81
Bio-Glas 500 (I)	20-50	34	0.50	4.2×10^{-5}	15
Bio-Glas 500 (II)	25-55	36	0.51	8.2×10^{-5}	30

a. From mercury penetration data. Electron miscroscopy (16) indicates wider distributions.
b. From mercury penetration data, estimated at half penetration in the internal pores.
c. From mercury penetration data.
d. TDPB for Bio-Glas 200 (I), and HDPB for the others.
e. Assuming 0.60 nm^2/ion.

the data is maintained.

Sodium dodecyl sulfate (SDS) and sodium tetradecyl sulfate (STDS) were used as received from Mann Research Laboratories. Hexadecyl pyridinium bromide (HDPB) obtained from Eastman Organic chemicals was recrystallized twice from ethyl ether and then several times from distilled water. Tetradecyl pyridinium bromide (TDPB) was synthesized from 1-bromotetradecane (Eastman White Label) and pyridine (Eastman Spectro grade) by the method of Anacker and Ghose (17) by Dr. J. R. Cardinal of this laboratory.

The c.m.c. values of the surfactants used at 30-35°C in water are estimated to be 7.8×10^{-4} M for HDPB, 3.0×10^{-3} M for TDPB, 2.1×10^{-3} M for STDS and 8.3×10^{-3} M for SDS (18). In the presence of added salts the c.m.c. values are much lower (18).

Adsorption Experiments. All solutions were made in double distilled water by weight. Bio-Glas samples were conditioned by maintaining in hot distilled water for about three hours. Adsorption experiments were performed in constant temperature baths by adding 10-15 ml of surfactant solutions of known concentration by weight to small Erlenmeyer flasks or Teflon-capped vials containing usually 1 g of Bio-Glas and occasionally 2 g of Bio-Glas at high surfactant concentrations. The Bio-Glas to liquid ratio was usually maintained constant in a particular series of measurements. Gentle shaking was employed to prevent foaming. The anionic surfactants were analyzed after suitable dilutions by weight by using the method of extraction into chloroform as methylene blue salts (19) followed by absorbance measurements with a Cary spectrophotometer, Model 16. The cationic surfactants were analyzed by absorbance measurements of the pyridinium chromophoric system at 259 nm, after suitable dilution by weight.

The adsorption experiments were run over periods of one to several days to follow the time dependence. Usually the changes in adsorption observed after 1-2 days below the c.m.c. and 2-4 days above the c.m.c. were within the estimated experimental errors.

The amount of surfactant adsorbed was calculated from the amount lost from solution in the usual manner. All adsorption data reported are thus apparent adsorptions not corrected for co-ion or micellar exclusions.

The average reproducibility and self-consistency of the data in a particular series of measurements was of the order expected from uncertainties of about 1-2% of the analytical concentration determination of the surfactant.

Results

Figure 1 shows the adsorption isotherms of HDPB to two samples of Bio-Glas 500 below the c.m.c. The isotherms show the two-step characteristic often observed for surfactant adsorption (8,20). The isotherms for samples I and II show roughly the same shape and

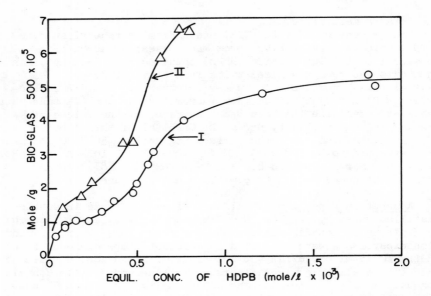

Figure 1.   Adsorption of HDPB to Bio-Glas 500, samples I and II, at low concentrations and 35°C

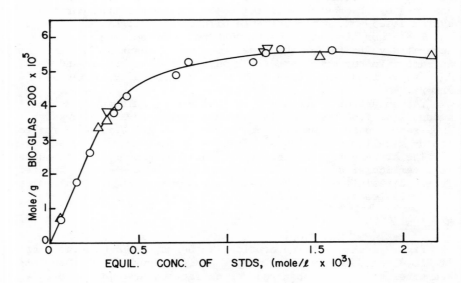

Figure 2.   Adsorption of STDS to Bio-Glas 200 (1) at low concentrations and 30°C.
△ = 1 day, ○ = 2 days, and ▽ = 3 days shaking.

indicate that the two samples have different surface areas. The isotherm for sample II shows a steep increase immediately below the c.m.c. region. The data at higher concentrations shows gentle increase above the c.m.c.

Figure 2 shows the adsorption isotherm of STDS below the c.m.c. The isotherm exhibits a plateau, or a slight maximum, well below the c.m.c. (2.1×10^{-3} M in water).

Figure 3 compares the adsorption isotherms of HDPB and TDPB above the c.m.c. for Bio-Glas 500 (II). The differences are small in the absence of added salt: for both surfactants the adsorption increases slowly above the c.m.c. In presence of high salt concentration, the apparent adsorption for HDPB decreases with concentration above the c.m.c.

Figure 4 shows the adsorption isotherms for TDPB for Bio-Glas 200 (I), above the c.m.c. from distilled water and in presence of 0.05M NaBr. It also shows the isotherm for HDPB for Bio-Glas 200 (II). The adsorption in all cases appears to be constant within experimental error above the c.m.c. 0.05M NaBr increases the adsorption above the c.m.c. by about 25% for TDPB as compared to the adsorption in the absence of salt.

Figure 5 shows the adsorption isotherm of STDS for Bio-Glas 200 (I) at high concentrations. The adsorption appears to decrease rapidly above the c.m.c. The effect of a 5°C change in temperature appears to be within experimental error.

Figure 6 compares the adsorption of STDS from water and from 0.01M and 0.03M NaCl for Bio-Glas 200 (I), obtained from the same bottle. The isotherm in the absence of salt is similar in shape to the 30° isotherm of Figure 5 and appears to be shifted upward by a constant amount of roughly 9×10^{-6} equiv./g. In presence of salt, the apparent adsorption remains nearly constant until the equilibrium STDS concentration reaches a value of about 10^{-2}M, above which it decreases.

Figure 6 also shows some adsorption data for SDS on Bio-Glas 200 (I) above the c.m.c. region. The apparent adsorption appears to be negative at high concentrations.

Discussion

pH Measurements. It is well known that glass is negatively charged at near neutral values of pH, and that H^+ and OH^- ions act as potential determining ions (21). The electrostatic factors involved in the adsorption of ionic surfactants are thus expected to be dependent upon the solution pH. The surfactants used in this work were salts of strong acids which themselves do not affect the pH of the solution, as was checked by measurements on our samples. In general, the adsorption of ionic surfactants to oppositely charged surfaces is expected to involve some ion exchange as part of the total uptake of surfactants (8,15,22). For cationic surfactants, this ion exchange with the glass surface is expected to release some H^+ ions into the solution (8,22). The interpretation

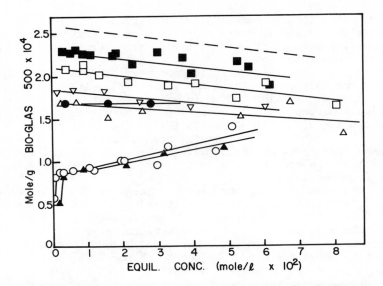

Figure 3. Adsorption of HDPB and TDPB to Bio-Glas 500 (II) at high concentrations and 35°C. ▲ = TDPB, ● = TDPB in 0.05M NaBr, ○ = HDPB, △ = HDPB in 0.03M NaBr, ▽ = HDPB on 0.05M NaBr, □ = HDPB in 0.13M NaBr, and ■ = HDPB in 0.2M $NaNO_3$. Dashed line indicates an exclusion volume of 0.5 ml/g.

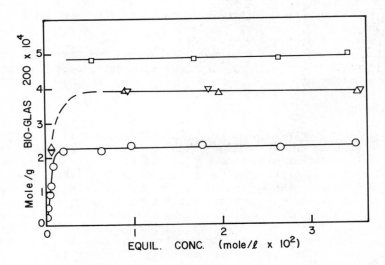

Figure 4. Adsorption of TDPB to Bio-Glas 200 (I) and HDPB to Bio-Glas 200 (II) at high concentrations and 35°C. ▽ = TDPB 4 days, △ = TDPB 7 days, □ = TDPB in 0.05M NaBr, and ○ = HDPB.

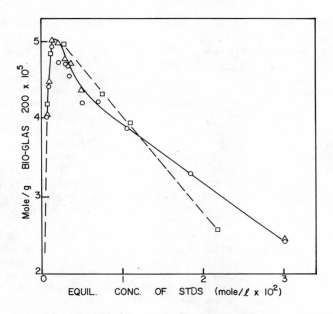

*Figure 5.   Adsorption of STDS to Bio-Glas 200 (I) at high
concentrations. ○ = 4 days, 30°C; △ = 6 days, 30°C;
□ = 7 days, 35°C.*

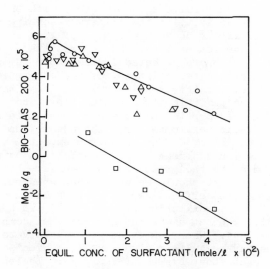

*Figure 6.   Adsorption of STDS and SDS to Bio-
Glas 200 (I) at high concentrations and 35°C.
○ = STDS, △ = STDS in 0.01M NaCl, ▽ =
STDS in 0.03M NaCl, and □ = SDS.*

of adsorption isotherms below the c.m.c. must take this factor
into account.

Some pH measurements were carried out to examine the nature of
the changes in solution pH brought about by the interactions of
the surfactants with Bio-Glas under the conditions of the adsorption experiments after equilibrium is attained. Figure 7 shows
some typical pH value of the supernatant liquid. The equilibrium
liquid was used to avoid the problem of the suspension effect on
pH (9,12,21). Bio-Glas itself, in the absence of any surfactant,
makes the medium slightly basic, indicating the release of small
amounts of basic electrolytes. Somewhat more pronounced effect is
obtained with Bio-Glas 200 than Bio-Blas 500. This is consistent
with the greater surface area of the former. When TDPB, a cationic surfactant, is added to Bio-Glas 500, some protons are released, as expected, and the solution pH becomes acidic. With increasing concentration of surfactant in the supernatant, however,
above the c.m.c., the pH changes very little, increasing by about
0.1 unit as the equilibrium concentration changes from the c.m.c.
to more than 15 x c.m.c. In the case of anionic surfactants which
are salts of strong acids, extensive ion exchanges are not expected. Figure 7 shows that in the STDS-Bio-Glas 200 system the
pH rises somewhat at low concentrations. The change is quite
small, however, and above the c.m.c., in particular, the pH
changes very little again, decreasing by about 0.1 unit as the
concentration increases to 7 x c.m.c., to about 1.5×10^{-2}M. A
further decrease of 0.2 units was observed when the equilibrium
concentration was 4×10^{-2}M, i.e. about 18 x c.m.c.

The pH variation results are consistent with the theoretical
argument that the activity of the primary adsorbing species, the
monomer, changes little above the c.m.c. of long-chain surfactants
and, therefore, the ion exchange and the resultant change in the
pH of the equilibrium are affected little above the c.m.c. The
reproducibility and self-consistency of the adsorption measurements below the c.m.c. (Figures 1 and 2) indicate that any change
in solution pH was sufficiently reproducible to not alter the
energetics of the adsorption process in a random fashion.

The c.m.c. values estimated from the adsorption isotherms
(Figures 1 and 6) do not differ much from expected bulk values,
showing that the ionic strength of the solutions is not affected
much at the c.m.c. by the presence of Bio-Glas. For comparisons
with other systems, it should be noted that when surfactants are
weak electrolytes, pH effects can be complex because of composition changes in the equilibrium solution (23) and because the pH
may be very different at a charged surface when compared to the
solution value (24). When ion exchange is likely to occur, initial pH values may be quite misleading, as indicated by the pH
variations brought about by TDPB (Figure 7). The pH may, indeed,
be difficult to control without buffers of reasonable capacity.
The measurement of the pH should be carried out on the equilibrium
fluid to avoid the suspension effect which can be serious (9,12,

21).

For the interpretation of the adsorption data above the
c.m.c., our reference point is always an internal one, namely the
adsorption at the c.m.c. of the same system. The above results
and the supporting arguments suggest that pH variations in solutions above the c.m.c. with respect to the solution at the c.m.c.,
and the effects of these changes on the adsorption, can be ignored
to a good approximation.

Tadros (25), in a recent paper, has reported that the adsorptions of a cationic and an ionic surfactant on silica at high
concentrations, but well below the c.m.c., change by less than a
factor of five as the pH changes from 3.6 to 9.1 for the cationic
system, and 3.6 to 10.1 for the anionic system. This effect
appears to be rather small but supports our neglect of pH-varia-
tion effects above the c.m.c.

Surface Areas. Tamamushi and Tamaki (8,26) found that the
limiting adsorption of dodecyl pyridinium bromide on alumina
was much less than that of dodecyl and higher alkylammonium chlorides, the limiting area being 0.90 nm^2/molecule. In our case, the
effect of electrolytes (Figure 3) indicates that the adsorption at
the c.m.c. on Bio-Glas in the absence of electrolytes is much less
than the maximum possible value. For an approximate estimate of
the effective surface area of Bio-Glas, we have used the value of
0.60 nm^2/molecule and the adsorption of the cationic surfactants
at the c.m.c. in the absence of added salt, which occurs from an
equilibrium pH of about 6. Table I reports these estimated surface areas.

Adsorption at and Below the C.m.c.

Cationic Surfactants. The adsorption isotherms for cationic
surfactants below the c.m.c. (Figure 1) are similar in shape to
those recorded for a variety of substrates (8,20,26). The net
uptake of the surfactant involves ion exchange to some extent,
as revealed by the lowering of the pH (Figure 7). The shape of
the isotherms indicates contributions from multilayer formation
(8,26,27) and/or attractive lateral interactions in the adsorbed
layer between the hydrocarbon groups (28).

Figure 3 shows that HDPB and TDPB appear to be quite similar
in their ability to adsorb at their respective c.m.c. values. It
has been noted before that the adsorption isotherms of homologous
long-chain ionic surfactants are often very similar if a reduced
concentration scale is used, i.e. if the concentrations are expressed as fractions of the c.m.c. (8,26,29). This corresponding
state type of correlation is of great interest but it poses a
problem with respect to surface coverage since different homologues are expected to occupy different surface areas, at least
for the fraction of the adsorbed molecules which lie parallel to
the surface. Shorter chain homologues should thus adsorb more on

the basis of smaller molecular surface areas. On the other hand,
lateral interactions may increase nonlinearly with chain length.
For ionic surfactants, however, two corrections may be of some
importance. From previous electrokinetic studies $(\underline{21},\underline{28},\underline{29},\underline{30})$,
it is extremely likely that there is a charge reversal well below
the c.m.c. For the diffuse double layer, therefore, the surfactant is usually a co-ion when the equilibrium concentration is
near the c.m.c. and the negative adsorption of the surfactant co-
ions, which is a function of the ionic strength of the system, ma
be somewhat greater for shorter-chain homologues with higher
c.m.c.'s.

The second factor involves the interaction (overlap) between
diffuse double layers surrounding adsorbed surfaces. Near the
c.m.c., this would result in an increase in the electrical work
opposing the adsorption of ionic species to similarly charged sur
faces and the adsorption, therefore, will be reduced. There may
be a small reduction of the exclusion of co-ions also $(\underline{31})$. In
our case, for Bio-Glas 500 (II), the ratio of the average pore
radius to the double layer thickness, δ, at the c.m.c. is about
1.8 for HDPB and 3.6 for TDPB. These ratios are such that considerable interactions between the double layers inside the pores
are expected at the c.m.c., particularly for HDPB. It is difficult, however, to make any quantitative calculations of the
effect.

In presence of neutral electrolytes at high concentration, the
adsorption of the cationic surfactants at the c.m.c. increases
appreciably, in accord with the expected reduction of the repulsive electrostatic interactions involved and the removal of double
layer overlaps. When the ionic strength is 0.05, for example, δ
is only about 1.4 nm. Figures 3 and 4 show adsorption data at
high ionic strengths for TDPB on Bio-Glas 500 (II) and Bio-Glas
200 (II) and HDPB on Bio-Glas 500 (II). At these high ionic
strengths the c.m.c. values are very low, less than $1 \times 10^{-4}M$ for
HDPB, and the adsorptions at the c.m.c. in Figures 3 and 4 can be
obtained by extrapolation of the post-c.m.c. lines to effectively
zero concentration. Figure 3 shows that the adsorption of HDPB in
presence of added NaBr increases with the concentration of NaBr.
The highest adsorption was observed for 0.2M $NaNO_3$ solution.

Anionic Surfactants. Unlike the case of cationic surfactants
(Figure 1), the adsorption of STDS to Bio-Glas 200 (I) below the
c.m.c. appears to attain a plateau value (or a very shallow maximum) well below the c.m.c. (Figure 2). The maximum adsorption is
only about 14% of that registered by the cationic TDPB, containing the same chain length as STDS. This is probably due primarily
to the unfavorable electrical interactions of anionic surfactants
on negatively charted surfaces.

For an estimate of the correction for co-ion exclusion for
STDS, it is to be noted that as the solution pH remains high in
presence of STDS (Figure 7), the glass surface should have a high

negative surface potential irrespective of the amount of STDS absorbed, and the surfactant anions are likely to be the predominant
co-ions of the electrical double layer. Co-ion exclusion can be
represented by an equivalent average distance, d, from the surface
up to which all co-ions are assumed to be excluded and beyond
which there is no exclusion (31). For planar noninteracting
double layers, when the surface potential is high and all mobile
ions are monovalent, this exclusion distance is represented very
well by the simple formula 2δ (31). The amount of exclusion per
cm^2 of surface is given by $2\delta C \times 10^{-3}$ where C is the bulk concentration of the co-ion in equivalents per liter. The amount is
thus approximately $6 \times 10^{-11}\sqrt{C}$ in water. If the surface area of
$140 \ m^2/g$ for Bio-Glas 200 (I) (Table I) is used, the exclusion is
calculated to be 8×10^{-7}, 2.4×10^{-6} and 3.7×10^{-6} equiv/g for
equilibrium concentrations of 10^{-4}, 10^{-3} and 2×10^{-3} equiv/l.
These calculations are of only approximate validity because of
double layer overlap effects but they indicate that the corrections may be significant.

In contrast to the comparable adsorption of HDPB and TDPB at
the c.m.c. of each, SDS, in the neighborhood of its c.m.c., seems
to adsorb much less than STDS. This may be due in part to reduced
chain-chain interactions in the adsorbed layers for the anionic
systems.

The adsorption of STDS in presence of 0.01M and 0.03M NaCl was
studied mainly above the c.m.c. (Figure 6). The extrapolated
values at the c.m.c. in presence of salt appears to be somewhat
less than the adsorption at c.m.c. in the absence of added salt.
Here also, the behavior of the anionic systems is at variance with
that observed for cationic systems (Figures 3 and 4). It is likely that the effective negative surface potential opposing anion
adsorption is controlled primarily by the solution pH, and does
not change much with concentration of NaCl.

Adsorption Isotherms above the C.m.c.

The complete interpretation of adsorption of ionic surfactants
above the c.m.c. must include (a) the effect of increasing monomer
activity above the c.m.c. (32,33,34), (b) the contribution of the
micelles to the ionic strength (24), (c) the effect of the changing composition of the solution on the electrical double layers
adjacent to the solid surface and their mutual interaction, as
also (d) the interaction of the micelles with the charged surfaces. Quantitative calculations of any of these factors appear
to be extremely difficult, and indeed, the factors may not be independent. In contrast to many studies of adsorption to finely
divided systems, the separation between adsorption surfaces in our
systems is independent of solution composition.

In the case of the anionic surfactants, particularly STDS
(Figure 2), the apparent adsorption, below the c.m.c., even after
reasonable corrections for co-ion exclusions, appears to change

relatively little over a concentration range of about a factor of
2. The effect of ionic strength variation on the adsorption at
the c.m.c. is also quite small (Figure 6). It seems, therefore,
that the effects of any further increase in monomer activity above
the c.m.c. and any increase in ionic strength and counterion concentration on the true adsorption can be ignored to a reasonable
approximation.

The striking decrease in the apparent adsorption of STDS and
SDS (Figures 5 and 6) above the c.m.c. can be qualitatively
rationalized in terms of exclusion of micelles from the surface.
The highly charged micelles can be reasonably expected to be re-
pelled by the negatively charged glass surface more extensively
than monovalent co-ions. Quantitative calculations are difficult
but some orders of magnitude can be estimated. As mentioned before, the equivalent exclusion distance for monovalent co-ions
from planar surfaces at high potentials is given by 2δ, when the
mobile ions are all monovalent and when interactions between
double layers is negligible. de Haan ($\underline{31}$) has shown that when bi-
valent and trivalent co-ions are introduced in tracer amounts in
such systems, the exclusion distance becomes 2.667δ and 3.067δ
respectively. For the highly charged micelles, any reasonable
extrapolation would indicate a much higher value. A value of 6δ
is used below for exploratory purposes.

The exclusion volume for noninteracting double layers is given
by the surface area times the exclusion distance. For Bio-Glas
200 (I), with the estimated surface area of 140 m^2/g the excluded
volume for SDS ($\delta \approx 3.3$ nm at the c.m.c.), for example, using 6δ as
the exclusion distance, is calculated to be 2.8 ml/g. For our
porous glass systems, most of the surface is associated with the
pores and, therefore, the exclusion effects should be lower when
compared to nonporous systems for which all adsorption surface is
externally located. An estimate of the exclusion volume can be
obtained from the experimental data by assuming that the apparent
decrease in the adsorption above the c.m.c. is entirely due to
micellar exclusion, and calculating this exclusion from the decrease in adsorption from that at the c.m.c. for a given equivalent concentration of micelles in the equilibrium solution.
Figure 5 indicates that for STDS the exclusion volume is between
1.5 to 1 ml/g. From the dimensions of the pores and the exclusion
distances ($\delta = 6.6$ nm at the c.m.c.), the minimum value for the
excluded volume should be of the order of the pore volume, 0.7
ml/g. The experimental values all appear to be higher. For SDS,
also, about 1 ml/g of excluded volume is indicated by the data
(Figure 6).

In presence of NaCl, the exclusion effect for the STDS system
near the c.m.c. seems to be rather low although the apparent adsorption does appear to decrease from about 10^{-2} equiv/l of the
equilibrium concentration of STDS. The reasons for this complicated behavior are not clear. It should be noted that the surface
of glass is probably covered by a gel-like layer ($\underline{35}$), the effect

of which is difficult to estimate.

In the case of the cationic surfactants, the micellar exclusion effect is expected to be relatively much smaller when compared to anionic systems because of the far greater surface coverage and possibly lower surface potentials of the positively
charged surface at the c.m.c. Thus, for example, the decrease in
the apparent adsorption observed for STDS on Bio-Glas 200 (I) over
the concentration range of the c.m.c. to 3×10^{-2} equiv/l is about
3×10^{-5} equiv/g. For TDPB (Figure 4), this would correspond to a
decrease of only about 8% in the apparent adsorption. The apparent constancy of the adsorption of the cationics for Bio-Glas 200
(Figure 4) could thus be due to a small increase in the true adsorption above the c.m.c. to compensate for the micellar exclusion.

The increase in the true adsorption has to be fairly appreciable for Bio-Glas 500 (II) for both TDPB and HDPB (Figure 3) in the
absence of added electrolytes because the apparent adsorption itself seems to increase above the c.m.c. In the absence of detailed information about the external and internal surface, it is
difficult to estimate the proper excluded volume but for HDPB,
because of its low c.m.c., and high δ, this should be of the order
of the pore volume, 0.5 ml/g. In fact the HDPB adsorption appears
to increase above the c.m.c. with an average slope of about 0.9
ml/g. For this system, however, somewhat below the c.m.c. region
(Figure 1), the adsorption to Bio-Glas 500 (II) seems to increase
very rapidly, with a slope of about 15 ml/g, unlike the case of
STDS (Figure 2). A roughly ten-fold reduction in the slope above
the c.m.c. would account for the observed isotherm above the
c.m.c. Several factors such as the increase in the activity of
monomers above the c.m.c. as the concentration increases to more
than 60 x c.m.c. (33,36), as also the possible contribution of
the micelles to the ionic strength of the solution (24), could
contribute to this apparent increase, and the magnitude of this
effect does not seem to be excessive.

In presence of 0.05M NaBr, the apparent adsorption of TDPB to
Bio-Glas 500 (II) seems to be nearly constant above the c.m.c.
(Figure 3). At this high ionic strength, the excluded volume
interactions from the positively charged surfaces should be much
reduced. For the higher homologue, HDPB, however, the apparent
adsorption appears to decrease with concentration above the c.m.c.
in presence of high concentrations of electrolytes. Micellar exclusion in these systems cannot be ascribed to double layer
effects. Steric exclusion appears to be more likely. Anacker
and Ghose (17) have shown that very large asymmetric micelles
form in such systems. It seems that some of these giant micelles,
which are probably rod-shaped (37), and which show pronounced excluded volume interactions in solution (38), are excluded from the
pores because of steric reasons. The excluded volumes are of the
order of the pore volume of 0.5 ml/g, as can be seen by comparison of the isotherms with the line drawn to show this pore volume.

A recently proposed theoretical treatment indicates that these
very large micelles are polydisperse and their average degree of
aggregation increases rapidly with concentration (38). Chromatographic studies of such systems employing porous glass should be
of some interest.

Adsorption Maxima and the Exclusion of Micelles

The existence of maxima in adsorption isotherms of surfactants
above the c.m.c. appears to be anomalous from a thermodynamic
point of view because beyond the maximum the adsorption decreases
with increasing solute activity (14,15). Various attempts have
been made to rationalize these observed maxima in terms of impurities in the system (14) or possible reduction of the surfactant
uptake by solid surfaces by ion exchange above the c.m.c. (15).
The maxima observed in the present work for the anionic surfactants have been ascribed to micellar exclusion from similarly
charged surfaces causing a lower estimate of the apparent adsorption. The pronounced effects observed, resulting even in apparently negative adsorptions at high concentrations (Figure 6), are
undoubtedly due to the relatively low surface coverage of the anionic surfactants, and the presence of an initially charged surface. Nevertheless, micellar exclusion above the c.m.c. is expected to be a general phenomenon and the interpretation of all
adsorption isotherms above the c.m.c. must take this into account.
The question arises as to whether this phenomenon can provide a
general explanation for adsorption maxima observed in other
systems.

For solid substrates, adsorption maxima have often been observed in systems where the adsorption immediately above the
c.m.c. increased rapidly with concentration, i.e., with roughly
the same slope as below the c.m.c. (14,15,39) before decreasing
at higher concentrations. No evidence for such a rapid increase
above the c.m.c. has been observed in our work. Whether such increases are possible when finely divided solids have interacting
double layers with variable spacings and separations is not known.
The apparent reduction in the adsorption in several such systems
beyond the maximum seems to occur too rapidly also to be explained in terms of a simple picture of micellar exclusion. Thus,
for example, in the original work of Vold and Phansalkar (39),
which drew attention to the problem of adsorption maxima, the adsorption isotherm of SDS on carbon decreases above the maximum
with a slope of about 10^{-5} ml/cm^2 whereas a rough estimate based
on the 6δ exclusion distance employed earlier, using 3.3 nm for δ
at the c.m.c. of SDS, would be 2 x 10^{-6} ml/cm^2. Again, the possible complications of interacting solid particles at variable
distances of separation remain to be examined.

For some solid systems a gentle decrease in the apparent adsorption occurs over extended concentration ranges, 10 to 100
times the c.m.c. (15). The micellar exclusion effect appears to

be adequate for such systems, although other factors may be involved also.

Of considerable interest in this connection are some binding
studies of ionic surfactants with hydrophilic polymers (40,41) and
non-ionic surfactants (40). The apparent binding isotherms, obtained from dialysis equilibrium studies, appear to show maxima at
concentrations close to or somewhat above the c.m.c. for the
dialysate, when the retentate contains the nonpermeable polymer.

The interpretation of even such soluble systems is difficult
for the case of polymers because of uncertainties about the polymer dimensions. Some data (40) on the binding of hexadecyl pyridinium chloride (HDPC) to Tween 80, a non-ionic surfactant with a
bulky sorbitanpolyoxyethylene head group and having a low c.m.c.
of about 0.0013% (42), are thus of particular interest because
compact micelles are involved. The data were obtained from dialysis studies using nylon membranes which are impermeable to Tween
80 but are permeable to HDPC. Figure 8 shows the binding isotherm, the data points for which were read from a large scale
graph published by Deluca and Kostenbauder (40). The binding results in the formation of mixed micelles. It appears to increase
beyond the c.m.c. of HDPC, which was estimated to be 1.0×10^{-3}M
(40), before decreasing at higher concentrations. The increase in
binding immediately above the c.m.c. may be caused in part by the
presence of homologous impurities. The lower homologues are expected to bind less to the mixed micelles. Since about 70% of the
HDPC added is bound to Tween 80 at the c.m.c., the fraction of the
lower homologues may be considerably higher in the dialysate than
in the original surfactant, resulting in a higher apparent c.m.c.

The mean slope above the maximum indicates the high excluded
volume of about 160 ml/g of Tween 80. The mixed micelles have a
mole fraction of about 0.75 of HDPC at the maximum as estimated
from the uptake of HDPC, and using the estimated molecular weight
of 1400 for Tween 80, and are, therefore, highly charged. The
excluded volume has a value of about 91 ml/g of mixed micelle,
and about 240 ml/g of hydrocarbon core in the mixed micelle.

To examine if these extremely high excluded volumes are
reasonable in terms of the exclusion of micelles of HDPC from the
electrical double layers of the mixed micelles of HDPC and Tween
80, some comparisons with mutual exclusions of micelles in inter-
micellar interactions are revealing. The second virial coefficients of a series of alkyl sulfate micelles have been investigated (13). The virial coefficients have high values, that for
SDS, for example, being 43 times the value expected if the micelles, without their double layers, are treated as hard spheres.
The micelles with their electrical double layers thus behave like
equivalent hard spheres of much larger radii than those of the
micelles themselves. If these radii are calculated from the
appropriate molecular weights, densities and the second virial
coefficients, and expressed as $r + k\delta$ where r is the radius of
the micelle, δ is the double layer thickness, and k is a propor-

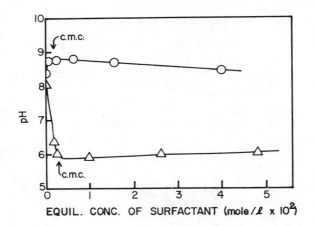

Figure 7. Variation of equilibrium solution pH with equilibrium concentrations at 30°C. ◯ = STDS with Bio-Glas 200 (I), △ = TDPB with Bio-Glas 500 (II).

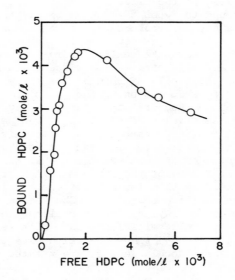

Figure 8. Adsorption (binding) isotherm of hexadecyl pyridinium chloride to 0.2% Tween 80 from dialysis equilibrium at 30°C. Data from reference (40).

tionality constant, so that the average equivalent distance between the micelle centers is $2(r + k\delta)$, the value of k is 1.0 for octyl sulfates, 1.25 for decyl sulfates and 1.34 for dodecyl sulfates at their respective c.m.c.'s. Thus the double layer contribution to the effective radius is of the order of δ or greater.

For an approximate calculation of the exclusion of HDPC micelles from the mixed micelles of HDPC and Tween 80 using an equivalent hard sphere treatment, we assume that both micelles have the same radii of the hydrocarbon core, 2 nm, and the double layer contribution to the effective radius, d´ in nm, is also the same. The volume around each mixed micelle excluded to HDPC micelle centers is given by that of a sphere of radius $2(2 + d´)$nm. Assuming the density of the micellar cores to be unity, the excluded volume of 240 ml/g of hydrocarbon core indicates that the ratio of $2(2 + d´)$ to 2 is $(240)^{1/3}$ so that the value of d´ is 4.2 nm. This value is appreciably less than the double layer thickness of about 10 nm at the c.m.c. of HDPC, and about 7.5 nm at the concentration corresponding to the maximum of the binding isotherm. Thus, the high exclusion volume observed for this system is actually less than what might be expected from the excluded volumes corresponding to the mutual interactions of the alkyl sulfate micelles close to the c.m.c. A number of factors may be responsible for this difference.

Exclusion of Polymers. The pronounced exclusions of co-ions and ionic micelles from charged surfaces discussed above are clearly the result of the long range of electrical interactions. For small uncharged solutes, such interactions are expected to be much weaker. For polyelectrolytes and even uncharged polymers, however, solute exclusion from adsorbed surfaces may be expected to be high, in view of the pronounced excluded volume interactions they exhibit in solution. The interpretation of their absorption isotherms would require an evaluation of this factor.

Acknowledgments

Partial financial support of this work by the General Research Support Grant by the Public Health Services is gratefully acknowledged. The authors thank Mr. Y. S. Yang for help with some of the experiments.

Summary

The problems associated with the interpretation of the adsorption and binding of ionic surfactants are reviewed and attention is drawn to the possibly important roles of co-ion exclusions and exclusions of micelles from similarly charged surfaces. The adsorption of two homologous cationic and anionic surfactants to porous glass (Bio-Glas) with average pore diameters of about 16 nm and 35 nm have been studied with special attention to equilibrium

concentrations above the c.m.c. The adsorption of the cationic
surfactants below the c.m.c. is associated with some ion exchange.
The adsorption of the anionic surfactants is considerably lower
than that of the cationics as expected from the negative charge on
the glass surface. Above the c.m.c. the apparent adsorption of
the cationics remains constant or increased slightly but that for
the anionics decreases and sometimes becomes negative at high concentrations. This decrease, and the associated adsorption maxima,
are attributed to the micellar exclusion effect and the plausibi-
lity of its magnitude is examined. The increase in the adsorption
of the cationics above the c.m.c. is shown to require small contributions from other factors such as increasing monomer activity
above the c.m.c. to compensate for the micellar exclusion effect.
The adsorption of the cationics is increased in presence of added
electrolytes. In systems in which large asymmetric micelles form,
an apparent reduction in the adsorption above the c.m.c. is observed in presence of high salt concentrations. It is attributed
to the exclusion of large micelles for steric reasons. The role
of micellar exclusion in the interpretation of adsorption isotherms of surfactants on solids as also binding isotherms to polymers and micelles, obtained from dialysis equilibrium studies, is
discussed. It is shown that the exclusion of micelles gives an
adequate explanation of the maxima observed in a binding isotherm
of cetyl pyridinium chloride above its c.m.c. to an impermeable
non-ionic detergent, Tween 80, in the micellar form. The possible
importance of solute exclusion effects for adsorption studies with
polymers and polyelectrolytes is indicated.

Literature Cited

1. Haller, W., Nature (1965), 206, 693.
2. Cantow, M. J. R., and Johnson, J. F., J. Polym. Sci., A-1
    (1967), 5, 2835.
3. van Oss, J., in "Progress in Separation and Purification,"
    E. S. Perry, Ed., Vol. 1, John Wiley, New York, 1968.
4. Jones, J. K. N., and Stoddart, J. F., Carbohyd. Res. (1968),
    8, 29.
5. Haller, W., J. Chromatog. (1968), 32, 676.
6. Cooper, A. R., and Johnson, J. F., J. Appl. Polym. Sci. (1969)
    13, 1487.
7. Hair, M. L. and Filbert, A. M., Research/Development (1969),
    20, 31.
8. Ginn, M. E., in "Cationic Surfactants," E. Jungermann, Ed.,
    Surfactant Science Series, Vol. 4, Marcel Dekker, New York,
    1970.
9. van Olphen, H., "An Introduction to Clay Colloid Chemistry,"
    John Wiley, New York, 1963.
10. Loeb, A. L., Overbeek, J. Th. G., and Wiersema, P. H., "The
    Electrical Double Layer Around a Spherical Particle," M. I. T.
    Press, Cambridge, Mass., 1961.

11. van Dolsen, K. M., and Vold, M. J., in "Adsorption from Aqueous Solution," Advances in Chemistry, Series No. 79, American Chemical Society, Washington, D. C., 1968.
12. Overbeek, J. Th. G., Prog. Biophys. Biophys. Chem. (1956), 6, 58.
13. Huisman, H. F., Proc. Konink. Ned. Akad. Wetenschap. (1964), B67, 407.
14. Moilliet, J. L., Collie, B., and Black, W., "Surface Activity," 2nd ed., van Nostrand, Princeton, N. J., 1961.
15. White, H. J., Jr., in "Cationic Surfactants," E. Jungermann, Ed., Surfactant Science Series, Vol. 4, Marcel Dekker, New York, N. Y., 1970.
16. Barrall, E. M., and Cain, J. H., J. Polym. Sci., Part C (1968), 21, 253.
17. Anacker, E. W., and Ghose, H. M., J. Am. Chem. Soc. (1968), 21, 253.
18. Mukerjee, P. and Mysels, K. J., "Critical Micelle Concentrations of Aqueous Surfactant Systems," NSRDS-NBS 36, Superintendent of Documents, Washington, D. C., 1971.
19. Mukerjee, P., Anal. Chem. (1956), 28, 870.
20. Tamamushi, B., in "Colloidal Surfactants," K. Shinoda, T. Nakagawa, B. Tamamushi, and T. Isemura, Eds., Academic Press, New York, N. Y., 1963.
21. Overbeek, J. Th. G., in "Colloid Science," H. R. Kruyt, Ed., Vol. I, Elsevier, New York, N. Y., 1952.
22. Ter Minassian-Saraga, L., J. Chim. Phys. (1960), 57, 10.
23. Smith, R. W., Trans AIME (1963), 226, 427.
24. Mukerjee, P. and Banerjee, K., J. Phys. Chem. (1964), 68, 3567.
25. Tadros, Th. F., J. Colloid Interface Sci. (1974), 46, 528.
26. Tamamushi, B. and Tamaki, K., Proc. Intern. Cong. Surf. Activity, 2nd, London, Vol. 3, p. 449, Academic Press, New York, N. Y., 1957.
27. Ter Minassian-Saraga, L., J. Chim. Phys. (1966), 2, 1278.
28. Somasundaran, P., Healy, T. W., and Fuerstenau, D. W., J. Phys. Chem. (1964), 68, 3562.
29. Benton, D. P., and Sparks, B. D., Trans. Faraday Soc. (1966).
30. Powney, J. and Wood, L. J., Trans. Faraday Soc. (1941), 37, 220.
31. de Haan, F. A. M., J. Phys. Chem. (1964), 68, 2970.
32. Mukerjee, P., Adv. Colloid Interface Sci. (1967), 1, 241.
33. Elworthy, P. H., and Mysels, K. J., J. Colloid Sci. (1966), 21, 331.
34. Abu-Hamdiyyah, M. and Mysels, K. J., J. Phys. Chem. (1967), 71, 418.
35. Tadros, Th. F., and Lyklema, J., J. Electroanal. Chem. (1969), 22, 9.
36. Mysels, K. J., J. Colloid. Sci. (1955), 10, 507.
37. Debye, P. and Anacker, E. W., J. Phys. Coll. Chem. (1951), 55, 644.

38. Mukerjee, P., J. Phys. Chem. (1972), 76, 565.
39. Vold, R. D. and Phansalkar, A. K., Rec. Trav. Chim. (1955), 74, 41.
40. Deluca, P. P. and Kostenbauder, H. B., J. Amer. Pharm. Assn., Scientific Edition (1960), 49, 430.
41. Lewis, K. L. and Robinson, C. P., J. Colloid Interface Sci. (1970), 32, 539.
42. Becher, P., in "Nonionic Surfactants," M. J. Schick, Ed., Surfactant Science Series, Vol. 1, p. 481, Marcel Dekker, New York, N. Y., 1967.

DTA Study of Water in Porous Glass

CHAUR-SUN LING and W. DROST-HANSEN

Laboratory for Water Research, Department of Chemistry, University of Miami, Coral Gables, Fla. 33124

Introduction

The properties of water near many interfaces are notably different from the properties of bulk water (or bulk aqueous solutions). Examples of unusual properties in interfacial behavior have been reviewed by various authors (1-4) and a few specific examples will be reviewed briefly in this paper together with some recent calorimetric measurements on water near silica surfaces.

The problem of the structure of bulk water remains unsolved; several excellent reviews of water structure have been presented recently (5-7). In view of the gap in our fundamental knowledge about water, it is hardly surprising that far less is known about the structure of water near interfaces. Nonetheless, some information is beginning to accumulate, suggesting various possible features of vicinal water structuring (3, 8-11). Specifically, it appears that more than one type of structure may be stabilized preferentially near an interface over a certain temperature interval, while other types of structures may prevail in other temperature intervals. Thus, relatively abrupt changes in properties of vicinal water have frequently been observed near the following temperature ranges: 14-16°; 29-32°; 44-46°; and 59-62°C. A review of the thermal anomalies and specifically, the role of this phenomenon in biological systems is given in (12). The occurrence of no less than four different transition regions suggests that there exists at least five different types of structures which may be stabilized adjacent to various interfaces. For this reason, it has been suggested previously (3, 9, 13) that the anomalies represent higher-order phase transitions in vicinally ordered structures. The purpose of the present study was to test this suggestion through a calorimetric study of the properties of water adjacent to silica surfaces.

Before proceeding, attention is called to the fact that water - especially in the solid state - may indeed occur in a

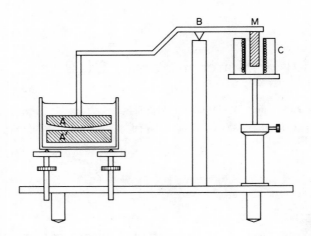

Zeitschrift für Physikalische Chemie

*Figure 1. Highly schematized drawing of disjoining
pressure apparatus by Peschel (17). B, Balance beam;
M, magnet; C, coil; A-A', optically polished quartz plates,
submersed in liquid under study.*

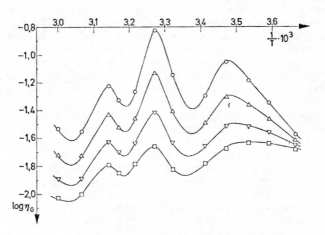

Naturwissenschaften

*Figure 2. Arrhenius plot of viscosity of water between
quartz plates (18) for various plate separations.* ○,
30 nm; △, *50 nm;* ▽, *70 nm;* ☐, *90 nm.*

large variety of structures. Thus, at least eight different,
stable high-pressure polymorphs exist in addition to an impres-
sive array of polyhedral clathrates (see particularly 12, pp.
55-60). It would hardly be surprising if somewhat similar structured elements might exist adjacent to an interface and be
transformed, at specific critical temperatures, from one type of
structure to another. The important question here is whether
the changes represent higher-order phase transitions or first-
order phase transitions. Obviously, a change from, say, Ice I
to Ice II, is a first-order phase transition, characterized by
a latent heat. If, instead, vicinally stabilized structures
transform from one structure to another without a specific latent
heat, the transitions will be second or higher-order (see 14-16).
    Measurements are reported in this paper on the thermal
effects observed in a DTA calorimetric study of water adjacent
to silica surfaces. The occurrence of structural changes at
discrete (and relatively invariant) temperatures are reasonably
well confirmed by the observed changes in slopes and baseline
shifts of the thermograms. In a number of cases, small endothermic peaks were observed in the thermograms (primarily seen
during heating), suggesting the occurrence of a latent heat of
transition. However, these peaks were by no means invariably
obtained (and in a few cases, apparent exothermic peaks were,
in fact, encountered). The utility of a calorimetry approach
to the problem of the nature of the thermal transitions in
vicinal water has been clearly demonstrated in this study, but
the experimental difficulties which were encountered prevent
any firm conclusions to be reached regarding the specific nature of the transitions, i.e., whether transitions are of
first or higher-order.

Background

    Before describing the results of the present investigation,
some pertinent earlier results from studies of vicinal water
will be reviewed.
    Peschel (17) and Peschel and Adlfinger (18, 19) have stu-
died the properties of liquids at interfaces using a highly
sensitive electric balance for measuring forces and displacements of two extremely carefully polished quartz places immersed in the liquid to be studied. Figure 1 shows a highly
schematized drawing of the device used by Peschel and Adlfinger.
With this instrument, it is possible to determine the effective, average viscosity of the water between the plates, as
the (exceedingly smooth) top plate "settles" onto the bottom
plate. Obviously, the rate of settling will depend on the viscosity of the liquid between the two plates. (Note that the
top plate is slightly curved; radius of curvature, one meter;
while the bottom plate is optically flat). The results of such
measurements are shown in Figure 2. Here the logarithm of the

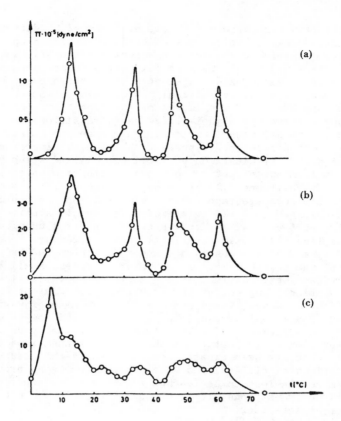

Figure 3.   Disjoining pressure for various plate separations: a) 50 nm; b) 30 nm; and c) 10 nm

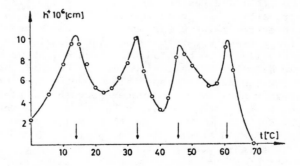

Figure 4.   Distance (h^+) for which the disjoining pressure is 10^4 dynes/cm² as a function of temperature

(apparent) viscosity is plotted as a function of temperature for
various plate separations. Note that distinct maxima and minima
are observed for all plate separations studied. This unusual
Arrhenius plot is one of the most astounding pieces of evidence
for the existence of abrupt changes in structure of aqueous interfacial layers, as discussed in the Introduction.

The apparatus has also been used by Peschel and Adlfinger
to measure the disjoining pressure between the two plates (19).
Some results from this study are shown in Figure 3. The observed disjoining pressure is seen to go through very sharp
maxima at a number of temperatures, namely, near 13, 32, 45 and
61°C (for a plate separation of 50 nm). From the disjoining
pressure data, Peschel and Adlfinger calculated the distance
at which the disjoining pressure attained an arbitrarily selected value (10^4 dynes per square centimeter). The result is
shown in Figure 4, from which it is seen that at the temperatures of the thermal anomalies, the distance at which the disjoining pressure reaches this particular value, approaches
0.1×10^{-4} cm (0.1μm).

The studies by Peschel and Adlfinger have added significant evidence for the existence of highly anomalous temperature-dependent properties of water near a quartz surface. Note
also that the distance over which the effects are observed is
of the order of 0.1μm - a distance which is extremely large
compared to "classical views" regarding structural effects near
an interface (such as implied in statements about "adsorbed
water", "surface modified water", "non-solvent water", etc.).
Distances of the order of 300 water molecule diameters are obviously of signal importance to such diverse aspects of aqueous
surface and colloid chemistry as nucleation rates, membrane
phenomena, and cell physiology, to mention but a few.

The measurements reported by Peschel and Adlfinger were
carried out only with pure water. However, thermal anomalies
near quartz interfaces have also been observed for both pure
water and aqueous solutions of electrolytes by Kerr and
Drost-Hansen (20). The instrument constructed by Kerr is a
long, thin-walled capillary, vibrating in an evacuated glass
cylinder. This device was originally developed by Thurn (21)
for measurements of viscosities (and more recently used for
measurements of densities (22).) The method was employed by
Forslind (23) to study the rheological properties of water. His
results are shown in Figure 5; the ordinate in this graph is the
half-life of the vibration times ($t_{1/2}$); the abscissa is the temperature. It is seen that a notable maximum in the half-life of
the vibrations occurs near 30°C.

The measurements by Kerr have confirmed the results obtained by Forslind. Figure 6 shows a typical run obtained by
Kerr, demonstrating an abrupt increase in the half-life of the
vibrations near 30°C. Similar results have also been obtained
with sodium chloride solutions in various concentrations (0.01;

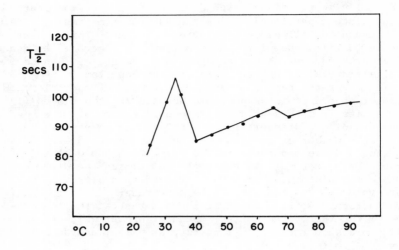

Svensk Naturv

Figure 5.   Half-life $(T_{1/2})$ in seconds, of vibrations of water-filled, "hairpin" capillary as a function of temperature (23)

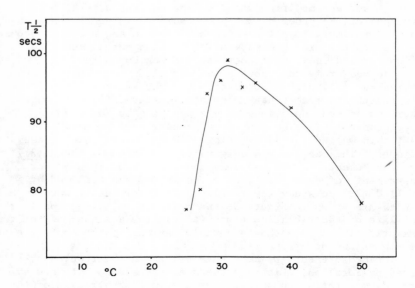

Figure 6.   Half-life $(T_{1/2})$, in seconds, of vibrations of water-filled, "hairpin"
capillary as a function of temperature (20)

0.05: 0.2 molar). However, for a 2 molar solution, the thermal
anomalies disappeared.

Figure 7 shows the resistance of the capillary (in the vibrating hairpin capillary viscometer) as a function of temperature. It is seen that near 30°C the temperature coefficient of
the resistance changes sign. The change in the resistance curve
is remarkably abrupt, similar to the changes in the half-life of
the vibrations. In a separate paper (10), one of us (W. Drost-
Hansen) has speculated on various molecular models to account
for the results obtained in this study and has compared these
findings to other studies of temperature-dependent surface properties.

As a final example of thermal anomalies, some recent data
by Wiggins (24) will be reviewed. Wiggins measured the adsorption of equimolar solutions of sodium and potassium salts on
silica gel. Equimolar solutions (0.075) molar) of either sodium chloride and potassium chloride, or Na^+ and K^+ iodide, or
Na^+ and K^+ sulfate were equilibrated with Davison silica gel,
type 950. Samples were "incubated" (with shaking) at various
temperatures overnight; after equilibration the supernatant
liquid was decanted from the silica gel. Analyses were then
made, using a flame photometer, of the concentrations of the Na
and K in the supernatant liquid and the liquid contained in the
pores of the silica gel. For each ion (sodium and potassium),
a partition coefficient, λ, was defined by the expressions:

$$\lambda_{K^+} = \frac{[K^+]i}{[K^+]o} \tag{1}$$

$$\lambda_{Na^+} = \frac{[Na^+]i}{[Na^+]o} \tag{2}$$

where the subscripts i and o, respectively, refer to pore solution and bulk solution. Using the two partition coefficients,
a selectivity coefficient, K_{Na}^{K} was formed as the ratio of the
two partition coefficients (Equation 3).

$$K_{Na}^{K} = \frac{\lambda_{K^+}}{\lambda_{Na^+}} \tag{3}$$

Figure 8 shows the temperature variation of the selectivity coefficient for the three salt combinations studied. It
is obvious that highly anomalous temperature dependencies
exist. For each pair of salts, maxima are obtained near 15,
30, 45 and 60°C, in amazingly good agreement with the temperatures proposed for thermal anomalies almost 20 years ago by one
of us (W. Drost-Hansen) (13) and the temperatures reported by
Peschel and Adlfinger (19). The curves shown in Figure 8 are
also seen to be almost superimposable (within experimental

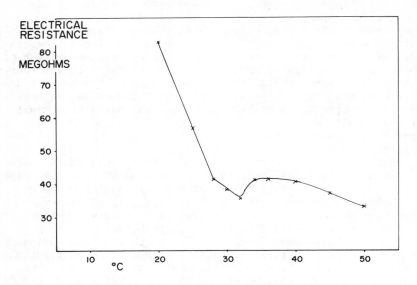

Figure 7. Electrical resistance (in megaohms) of 0.05M NaCl solution in "hairpin" capillary as a function of temperature (20)

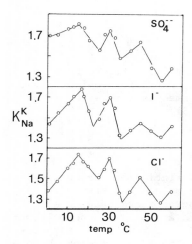

Biophysical Journal

Figure 8. Selectivity coefficient, K^K_{Na}, for potassium–sodium ion distribution near silica gel surface as a function of temperature (24)

errors) for the chlorides and iodides and qualitatively similar results are found for sulfates. This degree of consistency in the three sets of data, together with the rather small deviations for each point, makes it unlikely in the extreme that the observed temperature dependence is caused by some random experimental error. Note also that the values for the selectivity coefficient are all greater than one - no doubt a point of crucial importance in cellular biology where previously "active transport" was invariably invoked to explain the unusual distribution of potassium and sodium.

Experimental Procedures

A. Materials and Procedures. The silica used in the present study was obtained in the form of spherical, porous beads of a silicate glass with a network of interconnecting pores ("Bioglas"). The material is frequently used for gel permeation chromatography; various properties of the "Bioglas" porous material has been reported by Cooper, et al. (26). Samples were available with average pore diameters of 20, 100, 200 and 250 nm. The water used was obtained from a Millipore Super-Q ion exchange system.

Several methods were employed for studying the effects of the proximity of the water to the silica surface. A particularly useful approach consisted in comparing two samples with identical amounts of water and identical amounts of porous glass, but differing in pore diameters. This method will be referred to as the "double differential method".

The various techniques used in these experiments (to enhance any possible effect of the water near interfaces) are summarized below.

I.a) Equal amounts of porous glass (of identically same pore diameter) were weighed out in aluminum liners for the calorimeter cups, but different amounts of water were added (usually 4 microliter to one cup and 5 microliter to the other cup) on top of the dry materials (by means of a micropipet).

I.b) Samples of identical amounts and type of porous glass were placed in dessicators with different relative humidities (for at least two days). After an equilibrium state was reached (determined by successive weighings), samples of identical pore diameters were selected and compared with samples from dessicators with different relative humidities.

I.c) To the samples taken from the dessicators with different relative humidities, an additional amount of water (usually 3 microliter) was added to both sample and reference material.

II. The double differential method used employed identical amounts of porous Vycor material, but with different average pore diameters for sample and references. Normally the reference would be the largest diameter material (usually 250 nm). In

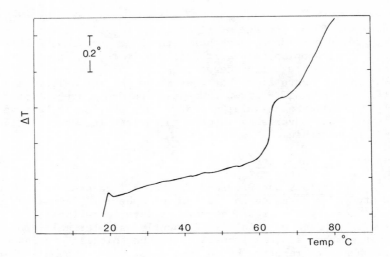

Figure 9.   Thermogram (Run #145) of water in porous glass; 10 mg porous glass beads plus 5 μl water in both cups. Pore diameters: reference, 250 nm; sample, 20 nm.

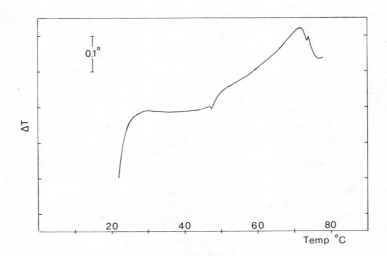

Figure 10.   Thermogram (Run #208) of water in porous glass beads. Pore diameter: 20 nm. Reference: 5 μl water; sample, 4 μl water; solid, 10 mg in each cup.

these cases, the following separate conditions were utilized:

II.a) Identical amounts of water were added to the samples
(usually 5 microliters). This gives a straightforward "double
differential" run.

II.b) Samples were placed in dessicators with the same relative humidity for at least two days. After an equilibrium
state had been reached, different average pore diameter samples
were selected. This gives a "double differential" set of data
with varying small amounts of (mostly adsorbed) water (due to
adsorption and, in part, capillary condensation).

II.c) Samples were used with different average pore diameters, maintained in a dessicator at identical relative humidities. To these samples an additional amount (usually 5 microliter) of water was added to the sample and reference material.

B. Instrumentation. A DuPont, Model 900, DTA system was
used with a calorimeter cell (DuPont, Model 900350). Various
heating rates and sensitivities have been employed. Generally,
the heating rates were of the order of 2° per minute (relatively slow); the sensitivities were usually (0.1 or) 0.2°
per inch (near maximum sensitivity).

One particular experimental difficulty has made quantitative estimates difficult or frequently impossible, namely,
slight asymmetric evaporation from the samples. To minimize the
evaporation, some runs were made using a "liquid-lid": approximately 5 microliter of dodecane was carefully pipetted into each
cup on top of the porous glass plus water samples. The addition
of dodecane is not expected to affect the overall thermal properties, although experimentally the situation has been complicated by the introduction of both a silica/dodecane and water/
dodecane interface. However, comparisons of experiments with
and without the organic liquid strongly suggest that the effect
of the dodecane is merely the (partial) suppression of evaporation effects, as intended.

Results and Discussion

The typical thermograms are shown in Figures 9 and 10. Experimental details and summary remarks of findings from representative runs are listed in Table I. Some additional comments are
appropriate: in runs in which different amounts of water have
been used, it is to be expected that a definite baseline slope
should be observed; however, the "baseline" is not straight,
but curved. This is probably due to different rates of evaporation of water from sample and reference cups; such evaporation
may possibly result in an exponential contribution to the baseline. (The initial part of the curve, approximately over a
3-5° interval, reflects an instability characteristic of the beginning of each thermogram, and this part of each curve was dis-
regarded).

Table I. Data from Thermograms; Water in Porous Glass Beads.

Run #	Pore diameter (nm) Sample	Ref.	Loading Sample	Ref.	Heating Rate °C/min.	Sensitivity °C/in.	Temp. °C	Features
299	20	20	52% RH + 3 μl	95% RH + 3 μl	2	0.1	48	sl. endothermic peak; change in slope
289	20	20	52% RH + 3 μl	95% RH + 3 μl	2	0.1	43	sl. endothermic peak; change in slope
184	20	20	4 μl	5 μl	0.5	0.1	32	sl. endothermic peak; change in slope
							43	abrupt offset in baseline
							58	change in slope
208	20	20	4 μl	5 μl	2	0.1	47	offset in baseline; change in slope
216	100	100	4 μl	5 μl	2	0.1	41	sl. change in slope
							59	offset in baseline; change in slope
218	100	250	5 μl	5 μl	2	0.1	52	change in slope
145	20	250	5 μl	5 μl	2	0.2	63	large offset in baseline (or exothermic peak?); change in slope

Table I. Data from Thermograms; Water in Porous Glass Beads. (Continued)

Run #	Pore diameter (nm)		Loading		Heating Rate	Sensitivity	Features	
	Sample	Ref.	Sample	Ref.	°C/min.	°C/in.	Temp. °C	
146	20	250	5 μl	5 μl	2	0.2	63	large offset in baseline
150	20	250	5 μl	5 μl	2	0.2	65	large offset in baseline
162	20	250	5 μl	5 μl	2	0.2	62	large offset in baseline
148	20	250	3 μl	3 μl	2	0.2	43-47	offset in baseline
151	20	250	3 μl	3 μl	2	0.2	49	abrupt offset in baseline
261	20	100	95% RH	95% RH	2	0.5	47;50	abrupt changes in slope
262	20	100	95% RH	95% RH	2	0.5	46;49	abrupt changes in slope
306	100	250	98% RH	98% RH	2	0.1	45	inflection point

The occurrence of small (apparent) endothermic peaks, and particularly changes in slope are characteristic of the results persistently obtained in the present study. Because of non-linear baseline slope, some of the data from these experiments have been replotted on a logarithmic scale, $\log(\triangle T)$ vs. T . No theoretical justification can be made for this procedure. However, if indeed the curvature is related to an exponential term in temperature, it is reasonable that taking logarithm of $\triangle T$ may remove some of the incurvation. This has been observed in many instances.

Sometimes very abrupt decreases occurred in $\triangle T$ near 70° to 80°C; this effect is probably an artifact due to rapid, asymmetric evaporation.

In all, more than 350 runs have been made. The results of some of these are summarized in the various tables. In Table II are listed the temperatures at which changes in slope and/or offsets in baseline and/or endothermic peaks have been observed. The material used in all of these runs had an average pore diameter of 20 nm.

Table III shows similar results obtained on identical amounts of porous glass with an average pore diameter of 100 nm. Again, the temperatures listed are those at which thermal anomalies were observed, such as changes in slope, offset in baseline, or endothermic peaks.

Table IV shows similar data for material with a 250 nm average pore diameter.

Table V lists results obtained by the double differential method, using 250 nm average pore diameter material as reference and 20 nm material as sample.

Table VI is similar to Table V, but for 100 nm vs. 20 nm.

Table VII is similar to table V, but for 250 nm vs. 100 nm.

Table VIII summarizes the results from the combined runs shown in Tables I through V. The average temperatures of transitions for "regions" II, III and IV agree well with the temperatures previously proposed for the transitions in vicinal water by one of us (W. Drost-Hansen). An anomaly near 74°C (Region V) appears not to have been reported before. Note that a number of anomalies were observed at temperatures where thermal anomalies have not previously been reported. The origin of these apparent anomalies is by no means clear, but it is conceivable that the "spurious anomalies" (at temperatures other than those generally encountered) may be related to the anomalies shown in the study by Peschel and Adlfinger (19) for very thin films of water (10 nm plate separation).

It is unfortunate that large, and frequently asymmetric, evaporation effects are encountered, as these appear to influence significantly all the thermograms obtained. However, in spite of the obvious experimental difficulties, it is certain from these measurements that rather abrupt changes occur in the thermal properties of the water/silica systems. The observed

Table II. Data from Thermograms of Identical Amount of Porous Glass Material (with an average pore diameter of 20 mm in reference and sample holders). Various Water Contents.

System	Run Number	Region I 12–18°C	Region II 27–34°C	Region III 42–48°C	Region IV 57–64°C	Region V 71–77°C	Others (°C)
5 μl (reference) 4 μl (sample)	177	–	33^k	–	63^k	–	–
	180	–	–	–	57^k	–	–
	183	–	–	–	–	–	53^k
	184	–	32^k	42^k	–	–	–
	208	–	–	47^k	–	74^k	–
	210	–	–	–	–	–	53^s
	211	–	–	–	–	–	56^k
	340	–	–	–	64^k	71.5^k	–
	295	–	–	43^s	–	–	–
	298	–	–	–	57^k	–	–
5 μl (reference) 4 μl (sample)	179	–	–	–	–	–	41^k, 67^k
	212	–	33^k	–	–	–	56^k
5 μl (reference) 4 μl (sample)	378	–	–	–	–	–	40^s
	379	–	–	42^s	–	75^s	–
95% RH (reference) 52% RH (sample) (3 μl in both)	278	–	31^k	43^k	–	–	–
	288	–	–	47^k	–	–	–
	289	–	–	43^k	–	–	38^k
	299	–	–	–	–	–	–
	300	–	–	–	59^s	–	–
	321	–	–	–	–	–	–
	322	–	–	–	–	–	56^s
	323	–	–	–	–	–	67^s
	337	–	–	–	–	–	50^s

Table II. Data from Thermograms of Identical Amount of Porous Glass Material (with an average pore diameter of 20 nm in reference and sample holders). Various Water Contents (Continued).

System	Run Number	Region I 12-18°C	Region II 27-34°C	Region III 42-48°C	Region IV 57-64°C	Revion V 71-77°C	Others (°C)
98% RH (reference)	292	-	-	-	-	-	37k
78% RH (sample)	293	-	34k	-	-	-	-
100% RH (reference)	340	-	-	-	64s	71.5s	-
98% RH (sample)		-	-	-	-	-	-

Footnote: k: small endothermic (or, rarely, exothermic) peak.
s: change in slope or offset of baseline.

Table III. Data from Thermograms of Identical Amount of Porous Vycor Material (with an average pore diameter of 100 nm in sample and reference cup). Various Water Contents.

System	Run Number	Region I 12–18°C	Region II 27–34°C	Region III 42–48°C	Revion IV 57–64°C	Region V 71–77°C	Others (°C)
5 μl (reference)	216	–	–	–	59s	–	–
4 μl (sample)	217	–	–	–	62s	–	–
	237	–	–	45s	–	–	–
	296	–	–	–	–	74s	70s
	235	–	–	42s	–	–	–
10 μl (reference)	220	–	–	–	–	–	79s
9 μl (sample)							
100% RH (reference)	245	–	–	46s	–	–	–
78% RH (sample)	246	–	–	42s	–	–	–
	272	–	–	42s	–	–	–
95% RH (reference)	279	–	–	42s	–	–	–
52% RH (sample)	294	–	33s	–	–	–	52s
(3 μl in both)	297	–	–	–	–	–	38s
100% RH (reference)	310	–	28s	–	–	–	–
95% RH (sample)							
100% RH (reference)	335	–	–	46s	–	–	–
52% RH (sample)	328	–	–	45s	–	–	–

Table IV. Data from Thermograms of Identical Amount of Porous Vycor Material (with an average pore diameter of 250 nm in sample and reference cup). Various Water Contents.

Sample	Run Number	Region I 12-18°C	Region II 27-34°C	Region III 42-48°C	Region IV 57-64°C	Region V 71-77°C	Others (°C)
5 μl (reference)	190	-	-	42^s	-	-	-
4 μl (sample)	301	-	-	44^s	-	-	-
95% RH (reference)	273	-	-	42^s	-	-	-
52% RH							
98% RH (reference)	309	-	32^s	-	-	-	-
78% RH	325	-	-	45^s	-	-	-
	133	-	-	-	-	-	38^s

Table V. Data from Thermograms of Identical Amounts but Differing Average Pore Diameters of Samples (Reference, 250 nm; Sample, 20 nm).

System	Run Number	Region I 12-18°C	Region II 27-34°C	Region III 42-48°C	Region IV 57-64°C	Region V 71-77°C	Others (°C)
5 μl of water in both	139	–	–	–	–	–	66s
	140	–	–	–	60s	–	–
	143	–	–	–	64s	–	–
	145	–	–	–	60s	–	–
	146	–	–	–	64s	–	–
	150	–	–	–	64s	–	–
	156	–	–	–	–	75s	–
	158	–	–	–	–	–	68s
	160	–	–	–	–	74k	–
	161	–	–	–	–	75k	–
	162	–	–	–	61s	–	–
	163	–	–	–	–	–	70k
	164	–	–	–	–	–	67s
	215	–	–	–	64s	–	–
3 μl of water in both	148	–	–	43.5s	–	–	–
	149	–	–	–	–	–	38s
	151	–	–	–	–	–	49k
	138	–	–	46k	–	–	–

Table VI. Data from Thermograms of Identical Amounts but Differing Average Pore Diameters of Samples (Reference, 100 nm; Sample, 20 nm).

System	Run Number	Region I 12-18°C	Region II 27-34°C	Region III 42-48°C	Region IV 57-64°C	Region V 71-77°C	Others (°C)
5 µl of water in both	219	-	-	-	-	-	53[s],66[k]
	232	-	-	-	-	-	55[s]
	233	-	-	-	-	-	66[k]
95% RH (reference)	261	-	-	47[s]	-	-	-
95% RH (sample)	262	-	-	-	-	-	54[s]
100% RH (reference)	265	-	-	45.5[s]	-	-	-
100% RH (sample)		-	-	-	-	-	-

Table VII. Data from Thermograms of Identical Amounts but Differing Average Pore Diameters of Samples (Reference, 250 nm; Sample, 100 nm).

System	Run Number	Region I 12–18°C	Region II 27–34°C	Region III 27–34°C	Region IV 42–48°C	Region V 57–64°C	Others (°C)
5 μl of water in both	218	–	–	–	–	–	52s
	221	–	–	–	–	–	52s
95% RH (reference)	257	–	–	–	–	–	38s
95% RH (sample)	256	–	–	46s	–	–	–
98% RH (reference)	306	–	–	44s	–	–	–
98% RH (sample)		–	–	–	–	–	–

Table VIII. Average Temperatures of Thermal Anomalies Occurring in Regions II–V.

Region	Average Temperature (°C)	Standard Deviation (σ)	Number of Points	Number of Points in $\pm 1 \sigma$
II	32.0	$\pm$ 1.8	8	6
III	43.8	$\pm$ 2.2	25	19
IV	61.5	$\pm$ 2.5	15	13
V	73.8	$\pm$ 1.5	8	6

anomalies are mainly changes in slope, displacements of baseline, or (less frequently) small endothermic peaks (in heating
curves). The changes in slope and the changes in baseline will
be discussed first. As all the results were obtained with a
calorimeter cell, the changes reflect variations in specific
heat and/or heat conductivity. In either case, such changes are
most likely associated with changes in the state of water at the
silica/water interface. Most likely, the changes are due primarily to variations in heat capacity. However, whether the
changes are in the heat capacity or due to changes in thermal
conductivity, these changes must reflect structural phenomena.
Unusual changes in thermal conductivity of bulk water have been
reported by Frontasev (27). However, those changes are likely
spurious effects due to interfacial phenomena, as discussed by
one of us (9, 28, 29) rather than manifestations of a bulk property. Metzik and Aidanova (30) have discussed exceptional
thermal conductivities of water near interfaces. See also the
brief survey article by Metzik, et al., (31). Finally, note
that any change in structure is likely to result in changes in
both the specific heat and the thermal conductivity.

DTA has frequently been used for the study of anhydrous
lipids, aqueous lipid systems and various biologically important
membrane systems (32). Extremely complicated thermograms have
been reported for human erythrocyte membranes (33) and an interesting clinical application of DTA has been reported by Heydinger, et al. (34). Only rarely has differential thermal
analysis been used specifically to study the properties of interfacial water (see, however, Bulgin and Vinson) (35).

We observe in passing that thermal anomalies are found at
or near the same temperature regions in interfacial properties
of water adjacent to a large variety of materials, including
ionic, polar and non-polar solids. This surprising observation
has been referred to as the "paradoxial effect" (12, 36). The
role of adsorbed water on various hydrophilic surfaces has been
studied by Bernett and Zisman (37), who conclude that "the critical surface tension, as well as the surface energy of clean,
high-energy hydrophilic solids (like borosilicate glass, fused
quartz and crystalline α-alumina) after exposure to a humid
atmosphere is dependent upon the surface concentration of water
adsorbed on that surface, but that it is independent of the
chemical nature of the underlying solid". They further conclude
that this observation may likely be extended to any other high-
energy hydrophilic surface (metals and metal oxide).

An interesting review of the properties of water near
surfaces has been reported by Hartkopf and Karger (38), who have
studied aqueous interfacial properties by gas chromatography.
This study - while not necessarily generally consistent with the
findings by Drost-Hansen [compare particularly with Ref. (3)] - is
significant, as it provides an experimental approach to the important, but frequently neglected, area of study of layers of

water with thicknesses ranging from 3 nm to about 200 nm. Other techniques for the study of layer thicknesses in this region have been reported by Tschapek and Natale (39); see also the article by Adamson regarding studies of film thicknesses well exceeding monolayers (40).

The small endothermic peaks observed in some thermograms during heating present a problem in interpretation. Such peaks were frequently, but not invariably, observed. It is impossible to determine if these peaks are "real" or artifacts. Assuming the peaks are indeed real, and if all the water in the sample undergoes a first-order phase transition, the apparent heat of transition is between 10 to 100 cal/g mole water. While extremely small, such values are not entirely unreasonable. Recall that the lattice free energies between Ice I and Ice II is only 19 cal/g mole and that the heat of transition among any two adjacent high-pressure polymorphs in the phase diagram of water differ, at most, by 550 cal/g mole and frequently by only 200-300 cal/g mole (6, 12). One of us has already suggested that in view of the ease with which water molecules may be packed in vastly different structures, at relatively minor energetic expenses, it is reasonable to assume that many different types of vicinal water structure may also exist - each with its own thermal stability domain. On the other hand, the fact that the endothermic peaks are not invariably observed, as well as the general behavior of the (thermal) properties of vicinal water, suggest that the changes observed at the "critical temperature regions" may reflect higher-order phase transitions (and thus, not be expected to be associated with a heat of transition). A large change in specific heat may possibly give rise to the "apparent" heat of transition. Furthermore, although in principle straightforward, there appears to be some uncertainty as to the distinction between first and second-order phase transitions. See, for instance, the comments by Lieb (41) and by Mayer (42).

While thermal anomalies are reported frequently in the literature - where sufficiently closely-spaced measurements have been made on interfacial systems - notable exceptions have been encountered. See, for instance, the ultrasonic study reported by Younger, et al. (43). It is of interest to note that in the present experiments, anomalous thermograms have been obtained under very widely differing conditions. Thus, comparisons of samples of porous glass exposed merely to different relative humidities (say, 52% vs. 95%) have produced unmistakeable changes in slope. Qualitatively similar results are also obtained in studies where large amounts of water have been present (for instance, 5 microliters per 5 mg of solid). This suggests that the phenomenon is indeed an interfacial phenomenon (9), although the anomalies are certainly not primarily characteristic of a mono-molecular layer or pauci-molecular layer; compare the discussion of the depth of the vicinal water as estimated on the basis of the measurements by Peschel and Adlfinger (19). The study of the nature of the thermal anomalies is continuing in our laboratory by Differential Scanning Calorimetry (using sealed "cups" to eliminate the spurious evaporation effects).

Summary

The data obtained from a DTA calorimeter study of water in samples of porous glass beads ("Bioglas", average pore diameters 20, 100, 200 or 250 nm) suggest that the water in these systems undergoes relatively sharp transitions over certain narrow temperature intervals (notably near 30 to 33°C, 44 to 46°C and 59 to 52°C) in good agreement with earlier observations by Drost-Hansen (3, 8, 9, 10, 11), Peschel and Adlfinger (18, 19) and others (23, 24). Because of experimental difficulties (notably evaporation effects), it is not possible to determine with certainty if the observed transitions in the vicinal water are first-order phase transitions or higher-order phase transitions, as proposed earlier by one of us (9, 12). However, the data do confirm the existence of anomalous temperature dependencies in the thermal properties of interfacial water.

Acknowledgement

One of us (W. Drost-Hansen) wishes to thank Dr. S. A. Bach for introducing him to the use of DTA (calorimetric) measurements for the study of thermal properties of aqueous systems. The authors also wish to thank Professor Curtis R. Hare for helpful discussions; Mr. Lawrence Korson for his assistance in solving various experimental problems, and Ms. Lynda Weller for preparing the manuscript.

Literature Cited

1. Henniker, J. C. Rev. Mod. Phys., (1949), 21(2), 322.

2. Low, P. F. Advan. in Agronomy, (1961), 13, 269.

3. Drost-Hansen, W. Ind. Eng. Chem., (1969), 61(11), 10.

4. Deryagin, B. V., Ed. "Research in Surface Forces", Vol. 3, Consultants Bureau, New York, 1971.

5. Franks, Felix, Ed. "Water, A Comprehensive Treatise", Vols. 1,2,3,4,5, Plenum Press, New York, 1972, 1973, 1974.

6. Eisenberg, D. and Kauzmann. "The Structure and Properties of Water", Oxford University Press, New York, 1969.

7. Horne, R. A., Ed. "Water and Aqueous Solutions: Structure, Thermodynamics and Transport Processes", Wiley-Interscience, New York, 1972.

8. Drost-Hansen, W. Ind. Eng. Chem., (1965), 57(4). (Also published in "Chemistry and Physics of Interfaces". Symposium Proceedings, Sidney Ross, Ed., Am. Chem. Soc., Washington, D. C., (1965).

9. Drost-Hansen, W. Chem. Phys. Lett., (1969), 2, 647.

10. Drost-Hansen, W. J. Geophys. Res., (1972), 77(27), 5132.

11. Drost-Hansen, W. Soc. Exp. Biology, Symposium Proceedings, (1972), XXVI, 61.

12. Drost-Hansen, W. in "Chemistry of the Cell Interface", B, H. D. Brown, Ed., Chapter 6, pp. 1-184, Academic Press, New York, 1969.

13. Drost-Hansen, W. Naturwissen., (1956), 43, 512.

14. Temperley, H. N. V., "Changes of State", Cleaver-Hume Press Ltd., London, 1956.

15. Ubbelohde, A. R., "Melting and Crystal Structure", Clarendon Press, Oxford, 1965.

16. Stanley, H. E. "Introduction to Phase Transitions and Critical Phenomena", Oxford University Press, 1971.

17. Peschel, G. Zeit. Phys. Chemie, Neue Folge, (1968), 59, 27.

18. Peschel, G. and Adlfinger, K. H. Naturwissen., (1971), 56, 558.

19. Peschel, G. and Adlfinger. Zeit. F. Naturforsch., (1971), 26a, 707.

20. Kerr, J. E. D., "Relaxation Studies on Vicinal Water", unpublished Ph.D. Dissertation, University of Miami, Coral Gables, Florida, 1970.

21. Thurn, H., Materialprüf., (1963), 5(3), 114.

22. Kratky, O., Leopold, H. and Stabinger, H. Zeit. Angew. Physik., (1969), 27, 273.

23. Forslind, E. Svensk Naturv., (1966), 2, 9.

24. Wiggins, P. M., "Thermal Anomalies in Ion Distribution". Submitted for publication, 1974; see also Biophys. J., (1973), 13, 385.

25. Wiggins, P. M.   In press, J. Theoret. Biol., (1974).

26. Cooper, A. R., Cain, J. H., Barrall, E. M., II and Johnson, J. F., Separ. Sci., (1970), 5, 787.

27. Frontas'ev, V. P., Dokl. Akad. Nauk SSSR, (1956), III(5), 1014.

28. Drost-Hansen, W. in "Equilibrium Concepts in Natural Water Systems", Adv. Chem. Ser., No. 67, pp. 70-120, American Chemical Society, Washington, D. C., 1967.

29. Drost-Hansen, W. Proceedings First International Symposium on Water Desalination, Vol. I, pp. 382-406, U. S. Government Printing Office, 1967.

30. Metzik, M. S. and Aidanova, O. S. in "Research in Surface Forces", Vol. 2, B. V. Deryagin, Ed., pp. 169-180, Consultants Bureau, New York, 1966. See also ibid., Vol. 3, pp. 34-35, 1971.

31. Metzik, M. S., Perevertaev, V. D., Liopo, V. A., Timoschtchenko, G. T. and Kiselev, A. B. J. Colloid Interface Sci., (1973), 43, 662.

32. Oldfield, E. and Chapman, D. Fed. Am. Soc. Exp. Biol. Letters, (1972), 23(3), 285.

33. Jackson, W. M., Kostyla, J., Nordin, J. H. and Brandts, J. F. Biochem., (1973), 12(19), 3662.

34. Heydinger, D. K., Hammer, E. J., Pgeil, R. W. and Taylor, P. H. J. Lab. Clin. Med., (1971), 77(3), 451.

35. Bulgin, J. J. and Vinson, L. J. Biochimica Biophysica Acta, (1967), 136, 551.

36. Drost-Hansen, W. Ann. N. Y. Acad. Sci., (1973), 204, 100.

37. Bernett, K. and Zisman, W. A. J. Colloid Interface Sci., (1969), 29(3), 413.

38. Hartkopf, A. and Karger, B. L. Accounts Chem. Res., (1973), 6, 209.

39. Tschapek, M., Natale, I. and Santamaria, R. Kolloid-Z., Z. Polym., (1965), 211, 143.

40. Adamson, A. J. Colloid Interface Sci., (1973), 44, 273.

41. Lieb, E. in "Phase Transitions", Solvay Institute, Pro-
     ceedings XIV, Conference on Chemistry at University of
     Brussels, pp. 14-15, Interscience Publishers, London, 1971.

42. Mayer, J. E. in "Phase Transitions", Solvay Institute,
     Proceedings XIV, Conference on Chemistry at University of
     Brussels, p. 15, Interscience Publishers, London, 1971.

43. Younger, P. R., Zimmerman, G. O., Chase, C. E. and Drost-
     Hansen, W. J. Chem. Phys., (1973), 58(7), 2675.

Monolayer Studies V. Styrene–Acrylic Acid Films on Aqueous Substrates

ERWIN SHEPPARD and NOUBAR TCHEUREKDJIAN

Chemical Research Department, S. C. Johnson and Son, Inc., Racine, Wis. 54303

Introduction

Low molecular weight styrene-acrylic acid copolymers of varying degrees of polymerization were spread as monolayers on water and various salt solutions at neutral and alkaline pH's.

The monolayer properties of high molecular weight styrene-acrylic acid systems were studied by Müller (1), who could not attain complete spreading and consequently monolayers were not formed. Monomolecular film behavior of other high molecular weight copolymers have been studied successfully. For example, Fowkes et al. (2) have reported on C8-C18 α-olefins-vinyl acetate systems, Labbauf and Zack (3) have reported on methyl methacrylate-styrene systems, and Hironaka et al. (4) have discussed films of methyl acrylate and n-butyl acrylate copolymers. Glazer (5) has reported on the spreading properties of ethoxylin resins with molecular weights from 400-2500 prepared by the reaction of epichlorohydrin and 4,4-dihydroxydiphenyl-propane. The adhesive properties of these resins were correlated with their interfacial spreading properties.

Experimental

Table I lists the styrene-acrylic acid, S/AA, copolymers studied, the number average molecular weights, $\overline{M}_n$, and the acid numbers or neutralization values defined as milli-equivalents of base needed to neutralize one gram of the copolymer. The theoretical neutralization value was 4.5 meq. The starting monomer mole ratio for the polymerization was 0.41 for acrylic acid and 0.59 for styrene. The initiator used was benzoyl peroxide. The molecular weights were determined by vapor phase osmometry using ethanol as the solvent.

Pressure-area isotherms at 22°C were determined by the Wilhelmy platinum hanging plate method with automatic recording of both the film pressure and the area (6). Minor modifications were incorporated in the design of Mauer's (7) analytical

Table I.   Description of Styrene/Acrylic Acid Copolymers

Sample	Initiator Level (mole %)	Meq. of base/g of copolymer	$\overline{M_n}$
1	0.5	4.89	--
2	1	4.32	2540
3	2	4.18	2400
4	3	4.12	2250
5	5	3.96	1750
6	6	3.96	1500
7	7	3.80	1280

balance for recording changes in weight keeping the sensing
plate nearly stationary. International Rectifier Type CS120V6
photocells were used in the Wheatstone bridge circuit, miniatur-
ized equivalent tubes in the amplifier circuit, and a 75 V D.C.
power supply.

The film trough and barriers used to clean the substrate
surface and compress the monolayers were FEP Teflon-coated tool
aluminum. The trough was rinsed with deionized water before
each determination. Deionized distilled water and materials
of highest purity available were used throughout. The monolayers were spread from 10% ethanol and 90% benzene (w/w) solutions and were formed on deionized distilled water, on substrates adjusted to pH=9, on dilute solutions containing various
cations and on their hydroxides. About 15 minutes were allowed
for the spread films to equilibrate before compression was
started at a constant rate of 150 cm^2/mg/min.

Gaines (8) has summarized most of the pertinent experimental techniques and precautions to be used in measuring monolayer properties.

Results and Discussion

Terminology. Monolayers of styrene/acrylic acid low molecular weight copolymers were studied in order to determine the
interfacial properties among the several samples of different
molecular weights and acid values and to determine the effect
of pH and of various cations on their spreading behavior at
air/aqueous solution interfaces.

The specific area, A_0, at zero film pressure was obtained
by extrapolating the linear portion of pressure-area isotherm
to the area axis as shown in Figure 1B. In this study atypical
π-A curves as represented in Figure 1A were obtained for the
S/AA films under certain conditions. The pressures at which the
isotherms deviated from linearity in the direction of higher
compressibilities, $-A^{-1}$ $(dA/d\pi)_T$, were designated as π_{CE},
collapse pressure for the expanded region, and π_{CC}, collapse
pressure for the condensed region. The former represented the
initial collapse point and the maximum film pressure in going
from the higher areas to the transition region while the latter
represented the final collapse pressure and the highest pressure
attained prior to final collapse of the spread film. The des-
ignations for the various regions of the isotherm were made
purely for convenience in this discussion and to describe the
high and low area regions, and the transition area regions
(areas between A_{OE} and A_{OC}). Here A_{OE} represents the extrapolated limiting value of the high area region and A_{OC} represents the extrapolated limiting value of the low area region.

Spreading of S/AA Copolymers on Water. Hydrocarbon homopolymers such as polyethylene and polystyrene do not spread

completely on water and water soluble polymers such as poly-
acrylic acid give unstable films on neutral substrates (8). Thus
a proper balance of water soluble and water insoluble groups is
required in order to obtain stable, completely spread monolayers
of copolymers. In Figure 2 are shown three typical pressure-
area isotherms for the S/AA copolymers spread as monolayers at
the air-water interface. Each of the seven copolymers exhibited
the three distinct monolayer regions which are shown schematically in Figure 1A. They have a highly compressible intermediate
transition region as characterized by the $\triangle A_0 = A_{OE} - A_{OC}$ value
which is the magnitude of the transition region. Monolayer film
areas, collapse pressures, and $\triangle A_0$'s for the copolymers spread
on water are recorded in Table II.

A plot of $\triangle A_0$ vs. copolymer acid number is given in Figure
3. When the linear region of the $\triangle A_0$ vs. copolymer acid number
curve was extrapolated to $\triangle A_0 = 0$, the corresponding acid number
was found to be about 4.5. In other words, a copolymer of this
series with an acid number of at least 4.5 is required in order
to enhance spreading without intermediate collapse pressure and
transition regions. Copolymers with acid numbers less than 4.5
would have their characteristic partially collapsed transition
regions which divide the high and low film area regions of the
isotherms. It is fortuitous that the polymerization process
utilized a starting monomer composition with a theoretical neutralization value of 4.5 meq. which is numerically identical to
the extrapolated limiting value of 4.5 meq. for the spread films
when $\triangle A_0 = 0$. In spite of the hydrophilic nature of the S/AA
copolymers, collapse of the monolayers started at relatively high
film areas. The region over which collapse occurred is controlled
by the copolymer composition as stated above and shown in Figure
3. Therefore, the interfacial film properties of the copolymers
may be altered at will to give any desired surface intermolecular
cohesion and film-substrate adhesional properties.

The transition region, $\triangle A_0$, extended over larger monolayer
areas when the copolymer acid number was decreased. The lower
molecular weight S/AA copolymers were relatively more hydrophobic
when compared to the higher molecular weight samples as indicated
in Table I. As the surface film pressure is increased the non-
film forming, hydrophobic portions of the copolymers are assumed
to be squeezed out of the film carrying along some of their film
forming neighbors. Depending upon the structure of the copolymers, it appears that more of a polymer chain could be squeezed
out of the surface film with increasing polystyrene concentration in the copolymer. At high pressures, the composition and
structure of the copolymers appeared to dominate the A_{OC} values
while at lower pressures the strongly hydrophilic nature of the
acrylic acid groups appeared to control the spreading characteristics of the films without regard to the polystyrene content
as indicated by the almost constant A_{OE} values except for Sample
1 which appears to be a special case. These data are shown in

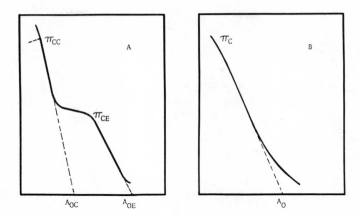

*Figure 1.   Schematic film pressure–area isotherms. Abscissa =
film area axis, ordinate = surface pressure axis.*

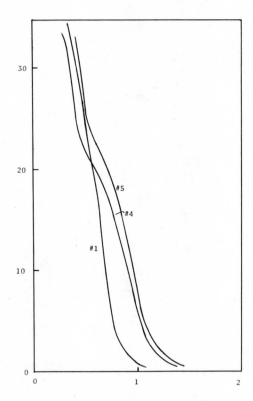

*Figure 2.   Pressure–area isotherms of styrene–
acrylic acid copolymers. Abscissa = area (m²/
mg), ordinate = film pressure (dynes/cm).*

Table II. Results of Spreading Styrene/Acrylic Acid Copolymers on Water

Sample	A_{OC} (m²/mg)	π_{CC} (dynes) cm	A_{OE} (m²/mg)	π_{CE} (dynes) cm	$\triangle A_O$ (m²/mg) ($A_{OE}-A_{OC}$)
1	0.77	35	0.82	16	0.05
2	0.90	32	1.04	15	0.14
3	0.89	33	1.16	16	0.27
4	0.72	31	1.13	15	0.41
5	0.74	33	1.14	16	0.40
6	0.73	31	1.21	13	0.48
7	0.56	32	1.10	13	0.54

Table II.   Thus, the structure of the high area film region may
be considerably different from the structure of the low area
region where the film exists in a partially collapsed state.   Because of the high rigidity of the copolymers, rearrangement of
the surface films to different orientations at high pressures
may not be probable.

   In the absence of a transition region ($\triangle A_0 = 0$), the structure of the film in the low pressure region could change gradually to the structure of the high pressure region.   With further
compression the film undergoes packing changes to minimize the
surface film pressure.   As a consequence, the film area decreases
by the elimination from the spread film of portions of the chain
not contributing significantly to film cohesion, substrate adhesion, and total energy of the system.   Therefore, the number of
substrate contact points would be reduced without causing major
structural changes such as chain folding.

   Detailed information such as copolymer configuration, end
group contributions, and solubility data are needed in order to
explain the small transition region exhibited by Sample 1 and
the variation of the individual A_{OC} values observed for each
film.   Glazer (5) found that for low molecular weight ethoxylin
resins the chain terminating group had a pronounced effect on
the monolayer properties and that isotherms of monolayers of
poly-γ-methyl-L-Glutamate spread on water had either α-helix or
β-extended chain structure depending upon the polypeptide configuration in the spreading solution and the composition of the
spreading solution (10).   Regardless of which copolymer was
examined, the isotherms showed hysteresis once they passed the
transition region and initial collapse point, π_{CE}.   The hysteresis loop was largest for Sample 7 and smallest for Sample 1;
however, they were not reproducible from one determination to
another.   The uniform compression rate even though probably too
fast for the formation of truly stable and fully extended structures nevertheless was sufficiently slow to give reversible
isotherms until π_{CE} was reached.

   Effect of Salts.   Monolayers of copolymers 2 and 7 were
formed on various aqueous salt solutions whose ionic strength
was adjusted to 8×10^{-4}.   Monolayer film areas and collapse pressures are tabulated in Table III.   The effect of adding soluble
salts to the substrate was to shift the low area region of the
π-A curves to higher areas probably because of film-substrate
interactions and thereby reduce the magnitude of the transition
region.   These changes in the film properties were independent
of the size and shape of the cations.   Moreover, the contribution of the various anions was indistinguishable from each other.
As expected, the changes in the monolayer properties were due
mainly to the cation charge.   In the presence of aluminum and
iron ions the transition region was eliminated.   The complexing
between the copolymer films and the trivalent cations and hydrated
cations apparently was strong enough to form a rigid structure

Table III. Limiting Areas and Collapse Pressures of Samples 2 and 7 Spread at Air/Solution Interfaces (ionic strength=8×10^{-4}, no pH adjustment)

Substrate	A_{OC} (m²/mg)	π_{CC} (dynes/cm)	A_{OE} (m²/mg)	π_{CE} (dynes/cm)	$\triangle A_O$ (m²/mg)
			Sample 2		
H_2O	0.90	32	1.04	15	0.14
Na^{+1}	0.94	29	1.08	14	0.14
Mg^{+2}	1.01	33	1.05	17	0.04
Ca^{+2}	0.97	34	1.08	18	0.11
Co^{+2}	0.95	30	1.06	15	0.11
Zn^{+2}	0.99	34	1.06	17	0.07
Al^{+3}	--	--	1.04	20	--
Fe^{+3}	--	--	1.03	27	--
			Sample 7		
H_2O	0.56	32	1.10	13	0.54
Na^{+1}	0.83	30	1.18	16	0.35
Mg^{+2}	0.71	35	1.15	14	0.44
Ca^{+2}	0.70	32	1.13	12	0.43
Co^{+2}	0.84	36	1.18	16	0.34
Zn^{+2}	0.76	35	1.18	16	0.42
Al^{+3}	--	--	1.08	24	--
Fe^{+3}	--	--	1.00	27	--

until final collapse.   In the case of univalent and divalent
cations, because of somewhat weaker interactions once again partial collapse, π_{CE}, occurred at relatively high areas.   However,
film folding was somewhat less than in the absence of cations as
mentioned above.

Effect of pH.   Monolayers of Samples 2, 4, 6, and 7 were
formed on water whose pH was adjusted to 9.   Table IV contains
the data for film areas, collapse pressures, and $\triangle A_0$ values.
Monolayers of Samples 2 and 7 were formed on aqueous substrates
adjusted to pH 9 which contained sodium, calcium, or zinc salts
at an ionic strength of 8×10^{-4}.   The limiting areas and collapse
pressures of Samples 2 and 7 spread on salt solutions are shown
in Table V.   The pH adjustments were made with sodium hydroxide.

Monolayers on aqueous solutions of pH 9.   The copolymers 2,
4, 6, and 7 are ionized completely at this pH.   At lower pH values, the S/AA monolayers will be a mixture of S/AA and S/AA$^-$.
The most significant effect on the monolayer properties was the
increase of the high pressure limiting area when compared to
films spread on neutral water.   The electrostatic repulsion of
the ionized acrylic acid groups apparently caused the increase of
the film area at low area regions.   At high area regions, the
ionization of the acrylic acid did not effect significantly the
spreading characteristics of the polymers probably because the
S/AA copolymers are relatively hydrophilic.

On complete ionization, the expansion at low areas would be
expected to be related to acrylic acid concentration in the
spread film.   Such a relationship is shown in Figure 4.   On the
average the films increased about 0.16 m^2/mg. of the copolymer or
0.54 m^2/mg. of acrylic acid.

The stability of the ionized films as indicated by the collapse pressures was comparable to the unionized films.   Apparently the decrease of the cohesiveness of the film due to ionization
was balanced by the increase in the interaction with the substrate.

Monolayers on aqueous salt solutions of pH 9.   Monolayers of
Samples 2 and 7 were formed on aqueous substrates at pH 9 containing Na^{+1}, Ca^{+2}, or Zn^{+2} ions.   For fatty acid molecules
spread on salt solutions, the role of the cations has been known
for a long time.   The formation of rigid salt structures at the
air-aqueous interface was considered to be the reason for the
changes in the π-A isotherms (8).

The effect of sodium salts on the π-A curves of S/AA$^-$ monolayers was nil when compared to monolayers on pH 9 water and consequently the isotherm was unaffected.   With calcium and zinc
ions in the substrate, the transition region was eliminated.   The
formation of diacrylates and condensation of the spread film appear to be distinct possibilities in the light of existing evidence of solution interaction between ionized acrylates and
cations to form various salts (11).   Apparently, when the completely ionized spread films interact with divalent ions, collapse

Table IV. Results of Spreading Styrene/Acrylic Acid Copolymers on Water Adjusted to pH=9

Sample	A_{OC} (m^2/mg)	π_{CC} (dynes) cm	A_{OE} (m^2/mg)	π_{CE} (dynes) cm	$\triangle A_O$ (m^2/mg) ($A_{OE}-A_{OC}$)
2	1.03	32	1.06	18	0.03
4	0.94	32	1.05	18	0.11
6	0.88	32	1.18	15	0.30
7	0.69	32	1.11	14	0.42

Table V. Limiting Areas and Collapse Pressures of Samples 2 and 7 Spread at Air/pH=9 Aqueous Solution Interfaces (ionic strength=8×10^{-4})

Substrate	A_{OC} (m²/mg)	π_{CC} (dynes/cm)	A_{OE} (m²/mg)	π_{CE} (dynes/cm)
		Sample 2		
H_2O	1.03	32	1.06	18
Na^{+1}	1.03	32	1.06	16
Ca^{+2}	0.98	36	--	--
Zn^{+2}	0.92	36	--	--
		Sample 7		
H_2O	0.69	32	1.11	14
Na^{+1}	0.70	33	1.09	14
Ca^{+2}	0.70	36	--	--
Zn^{+2}	0.89	34	--	--

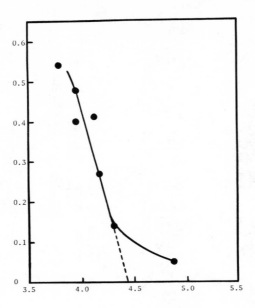

Figure 3.   The dependence of the transition region magnitude on the copolymer acid number. Abscissa = copolymer acid number, ordinate = A_0 (m^2/mg).

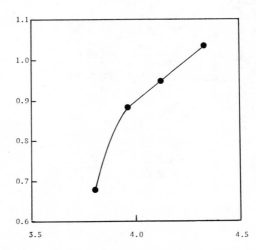

Figure 4.   Limiting film areas as a function of acid number at pH = 9. Abscissa = acid number, ordinate = A_{oc} (m^2/mg).

at higher areas does not occur and only one structure dominates throughout. This situation was similar to the films spread on aqueous Fe^{+3} and Al^{+3} salt solutions where the spread film was not completely ionized. The effect of the charge of trivalent cations apparently was strong enough to dominate the entire pressure range.

Conclusions

In spite of the acrylic acid content of the copolymers the monolayers of many of the styrene-acrylic acid copolymers spread somewhat poorly on deionized water.

Ionization of the acrylic acid groups aided in the spreading and stability of the spread films.

Trivalent cations promoted the spreading of the films at neutral pH.

Divalent cations promoted the spreading when the films were ionized at pH 9.0.

A minimum copolymer acid level for complete spreading may be determined using copolymers of varying compositions and extrapolating the linear region of the ΔA_0 vs. copolymer acid concentration to zero ΔA_0; and,

As expected, the interfacial film properties of the copolymers may be altered to give any desired π-A curves or surface properties.

Literature Cited

1. Müller, F. H., Z. Elektrochem. (1955), 59, 312
2. Fowkes, F. M., Schick, M. J., and Bond, A., J. Colloid Sci. (1960), 15, 531
3. Labbauf, A. and Zack, J. R., J. Colloid Interface Sci. (1971), 35, 569
4. Hironaka, S., Kubota, T., and Meguro, K., Bull. Chem. Soc. Jap. (1972), 45, (11), 3267
5. Glazer, J. J., J. Polym. Sci. (1954), 13, 355
6. Sheppard, E. and Tcheurekdjian, N., J. Colloid Interface Sci. (1968), 28, 481
7. Mauer, F. A., Rev. Sci. Instrum. (1954), 25, 589
8. Gaines, G. L., Jr., "Insoluble Monolayers at Liquid-Gas Interfaces," Interscience, New York, 1966
9. Crisp, D. J., J. Colloid Interface Sci. (1946), 1, 161
10. Loeb, G. I. and Baier, R. E., J. Colloid Interface Sci. (1968), 27, 38
11. Davidson, R. L. and Sittig, M., Ed., "Water Soluble Resins," 2nd ed., Reinhold Book Corp., New York, 1968.

11

Interaction of Calcium Ions with the Mixed Mono-layers of Stearic Acid and Stearyl Alcohol at pH 8.8

D. O. SHAH

Departments of Anesthesiology and Chemical Engineering, University of Florida, Gainesville, Fla. 32611

Introduction

Molecular interactions in mixed monolayers are of importance in understanding the phenomena such as stability of foams
and emulsions, retardation of evaporation by films, and
reactions occuring at the cell-surface (1-12). Earlier studies
on mixed monolayers of acids, amines, alcohols, ethers, and
triglycerides were reported by Schulman and his coworkers
(13-15). Similar studies were also reported by Harkins and his
group (16-18). Various investigators (19-29) have studied the
effect of di- and trivalent cations as well as the effect of pH
of the subsolution on the ionic structure of fatty acid monolayers. The present paper reports our studies on the ion-dipole
interaction between stearic acid and stearyl alcohol at pH 8.8,
as well as the interaction of calcium ions with these mixed
monolayers.

Materials and Methods

Highly purifies (>99%) stearic acid and stearyl alcohol
were purchased from Applied Science Laboratories, Inc., (State
College, Penn. 16801). Lipid solutions of 0.8 to 1.0mg/ml
concentration were prepared in methanol-chloroform-hexane
(1/1/3v/v/v) mixture, where all solvents were of spectroscopic
grade. Inorganic chemicals of reagent grade, and distilled--
deionized water of electrical resistance 1.2×10^6 ohms/cm were
used in all experiments.

The surface pressure was measured by a modified Wilhelmy
plate method, and the surface potential was determined by
using a radioactive electrode, as described previously (30).
The state of the monolayer was inferred from the mobility of
talc particles sprinkled on the monolayers when a gentle stream
of air was blown at the particles by means of a dropper (31).
Various states of monolayers were qualitatively distinguished
on the basis of movement of talc particles and designated as

170

the solid, gel, viscous liquid, or the liquid state. In the
solid state, the talc particles do not move at all; in the gel
state they move very little and stop; in the viscous liquid
state they move but not freely; whereas in the liquid state of
monolayers the particles move freely when the air current is
blown at them.

The surface measurements were taken on the following subsolutions: 0.05M tris buffer + 0.02M NaCl, and 0.05M tris
buffer + 0.01M $CaCl_2$ at pH 8.8. The buffer solutions were
prepared according to Biochemists' Handbook (32).

Theory

The average area per molecule in a mixed monolayer is
calcualted by dividing the total area by the total number of
molecules of both components in the mixed monolayer. If the
molecules of both components occupy the same molecular areas as
in their individual monolayers, the points for the average area
per molecule of the mixed monolayers would lie on a straight
line joining the two end points for the pure components at the
same state of compression. A deviation from this 'additivity
rule' indicates condensation of the mixed monolayers either due
to an <u>interaction</u> or <u>intermolecular cavity effect</u> in the mixed
monolayers (12,33).

Surface potential (ΔV) of a monolayer can be expressed as
$\Delta V = Kn\mu$, where K is a constant, n is the number of molecules
per cm^2 of the film (i.e., $n = 10^{16}$/area per molecule in $\mathring{A}^2$),
and μ is the resultant vertical component of the dipole moment
of the molecule. Thus, $\Delta V/n = K\mu$, where the term on the left-
hand side of the equation, representing the average potential
per molecule (mv/molecule), is proportional to the surface
dipole moment μ of the molecule. Hence, $\Delta V/n = \Delta V \times 10^{-16}$ x
area per molecule in $\mathring{A}^2$, where the term on the right-hand side
can be used for calculations.

The advantages of using $\Delta V/n$ instead of ΔV alone are the
following: firstly, the average potential per molecule ($\Delta V/n$)
is a parameter which is a characteristic of the molecule, and
is analogous to average area per molecule; and secondly, it
eliminates changes in surface potentials caused by expansion
or condensation of the mixed monolayers. Conclusive evidence
for ionic interaction can only be obtained from surface potential
measurements when $\Delta V/n$ is plotted against mole fraction of the
components in mixed monolayers at the same state of compression.
In this case a deviation from the additivity line indicates
ion-ion, or ion-dipole interaction between the two components
in mixed monolayers (12,33).

Results

Average molecular areas and potentials of the mixed

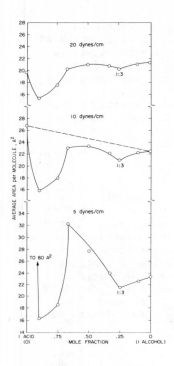

Figure 1. *Average area per molecule of stearic acid–stearyl alcohol monolayers at different surface pressures on subsolutions of 0.05M tris buffer + 0.02M NaCl, pH 8.8, at 22°C. The broken line indicates the additivity rule of molecular areas. The optimum condensation of mixed monolayers occurs at 9:1 and 1:3 molar ratios.*

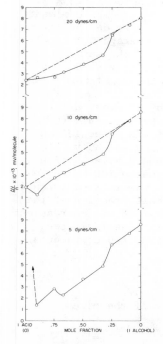

Figure 2. *Average potential per molecule of stearic acid–stearyl alcohol monolayers at different surface pressures on subsolutions of 0.05M tris buffer + 0.02M NaCl, pH 8.8, 22°C. The broken lines indicate the additivity rule of average potentials.*

monolayers of stearic acid and stearyl alcohol on subsolutions
of 0.05M tris buffer + 0.02M NaCl at pH 8.8 are shown in Figures
1 and 2.   Average molecular areas in the mixed monolayers show
optimum condensation at molar ratios 9:1 and 1:3 between stearic
acid and stearyl alcohol (Figure 1).   The average potentials
show deviations from the additivity rule at all surface
pressures (Figure 2).   The deviation is less prominent when
stearyl alcohol is the major fraction in the mixed monolayers.
    Figures 3 and 4 show average molecular areas and potentials
on subsolutions of tris buffer + 0.01M $CaCl_2$ at pH 8.8.   The
average areas follow the additivity rule (Figure 3), whereas
average potentials show deviation from it (Figure 4).   However,
in contrast to the nonlinear decrease in the absence of $CaCl_2$
(Figure 2), there is a linear increase in average potential in
the presence of $CaCl_2$ in subsolutions.
    Figure 5 shows the maximum values of surface potential of
mixed monolayers near their limiting areas on both subsolutions.
It should be noted that surface potentials of mixed monolayers
on tris buffer + NaCl subsolutions show distinct kinks at 9:1,
3:1, and 1:3 molar ratios of stearic acid to stearyl alcohol.
Table I summarizes the state of the monolayers near their
collapse pressures ($\approx$40dynes/cm) on various subsolutions.   The
mixed monolayers did not show significant differences in their
collapse pressure values.   The stearic acid monolayer was in
the solid state in the presence of $CaCl_2$.   However, the presence
of 10 mole % of stearyl alcohol converted the monolayer to the
gel state, suggesting a disording of solid lattice structure
of calcium stearate by stearyl alcohol.

Discussion

    At pH 8.8, the average molecular areas show two minimia at
molar ratios 9:1 and 1:3 between stearic acid and stearyl
alcohol in mixed monolayers (Figure 1).   We have shown (11)
previously that the rate of evaporation through these mixed
monolayers also show a minimum at the 1:3 molar ratio between
stearic acid and stearyl alcohol.   However, there was no such
minimum in the evaporation through mixed monolayers at the 9:1
molar ratio is presumably due to structural alterations (e.g.,
aggregation or micelle formation) in the mixed monolayer.
Using various mixed surfactant systems such as mixed monolayers,
foams and emulsions, we have proposed (11) that the striking
change in the properties of these systems at the 1:3 molar
ratio is due to a hexagonal arrangement of molecules of the
two components such that one component occupies the corners and
the other component occupies the center of a hexagon.   The
mechanism of structural alteration at the 9:1 molar ratio in the
mixed monolayers may be as follows.   At pH 9, it is expected
that there will be ionic repulsion between stearic acid molecules in the monolayer.   Perhaps a small number of stearyl

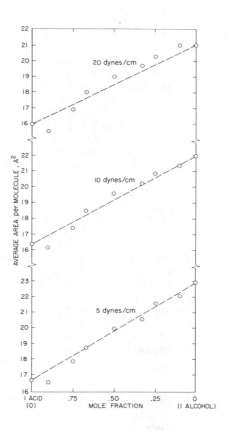

Figure 3. Average area per molecule of stearic acid–stearyl alcohol monolayers at different surface pressures on subsolutions of 0.05M tris buffer + 0.01M $CaCl_2$, pH 8.8, 22°C. The broken lines indicate the additivity rule of molecular areas.

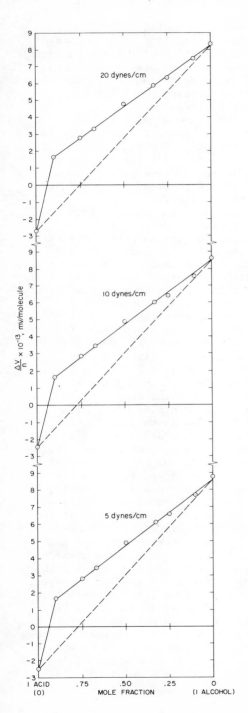

Figure 4. Average potential per molecule of stearic acid–stearyl alcohol monolayers at different surface pressures on subsolutions of 0.05M tris buffer + 0.01M CaCl₂ pH 8.8, 22°C. The broken lines indicate the additivity rule of average potential abruptly changes at a 9:1 molar ratio.

alcohol molecules act as "condensation nuclei" around which
ionized molecules gather to cause maximum interaction among
molecules. This may result into formation of three dimensional
micelles or aggregates within the mixed monolayer and a subsequent decrease in the average area per molecule at this ratio.
The minimum observed in the surface potential at 9:1 molar
ratio also supports this concept (Figure 5).

The average potentials of the mixed monolayers at pH 8.8
in the presence of NaCl are shown in Figure 2. The decrease
in the average potentials from the aditivity rule suggests an
ion-dipole interaction between components of the mixed monolayers (33). Similar decrease in the average potential due
to an ion-dipole interaction has been reported for monolayers
of dicetyl phosphate and cholesterol (12).

Figure 3 shows average molecular areas which follow the
additivity rule for the mixed monolayers at pH 8.8 in the
presence of $CaCl_2$. This suggests that the molecular association between stearic acid and stearyl alcohol at the 1:3 molar
ratio does not occur in the presence of $CaCl_2$ in the subsolution, and that the structural alteration that occurs at the
molar ratio of 9:1 also does not take place in the presence of
$CaCl_2$. The average potentials show a deviation from the
additivity rule (Figure 4). It is interesting that on both
subsolutions, the average potential showed a deviation from the
additivity rule (Figures 2 and 4). However, the deviation is
non-linear in the absence of $CaCl_2$ whereas linear in the
presence of $CaCl_2$. Moreover, in the absence of $CaCl_2$, the
ion-dipole interaction between the fatty acid and the alcohol
decreases the average potentials whereas interestingly the
average potential increases in the presence of $CaCl_2$ with
respect to the additivity rule (Figures 2 and 4). It is
presumably the ionic structure of calcium distearate that is
responsible for the increase in the average potentials. Our
proposed explanation for the interaction of calcium ions with
the mixed monolayer of stearic acid and stearyl alcohol is
shown schematically in Figure 6. It has been suggested (34,35)
that in stearic acid monolayers each calcium ion coordinates
with four carboxyl ions to produce a copolymeric soap lattice
which is shown in two dimensions in Figure 6(a). The copolymeric structure composed of ionized carboxyl groups and
calcium ions permits a maximum charge interaction in stearic
acid monolayers, and hence, gives a very low (negative) value
of average potential per molecule (Figures 4 and 5). However,
the presence of about 10 mole % or more of stearyl alcohol in
stearic acid monolayer causes considerable disruption of the
copolymeric structure, and hence, converts the copolymeric
structure into a calcium distearate form. Any further increase
in the fraction of mole % of stearyl alcohol simply dilutes
the calcium distearate units in the mixed monolayer. This
interpertation is supported by the results shown in Figure 4.

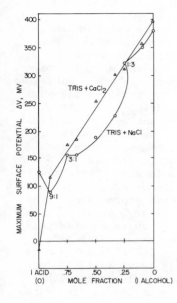

Figure 5. The maximum surface potential of stearic acid–stearyl alcohol mixed monolayers on various subsolutions at pH 8.8 in the presence (△) and absence (○) of CaCl₂. For subsolutions of 0.05M tris + 0.02M NaCl at pH 8.8, the kinks in the surface potentials occur at molar ratios 9:1, 3:1, and 1:3.

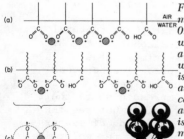

Figure 6. Schematic of stearic acid–stearyl alcohol monolayers on subsolutions of 0.05M tris buffer + 0.01M CaCl₂ at pH 8.8. (a): stearic acid monolayers in which nearly all carboxyl groups are ionized forming a copolymeric lattice where each calcium ion interacts with four adjacent carboxyl ions. One carboxyl group is located below and one above the diagram plane around a calcium ion. Therefore, all oxygen atoms of carboxyl groups are equivalent, as indicated by a solid and a broken line. The solid state of the monolayers is shown by the straight hydrocarbon chains. (b): mixed monolayers in which the stearyl alcohol molecule disrupts the copolymeric structure which exists in stearic acid monolayers, resulting in a formation of calcium distearate at pH 8.8, in which all four oxygens are equivalent (δ−) and interact with a calcium ion. The circles in broken lines indicate cross-sectional areas of stearic acid molecules.

It shows that the average potential varies linearly from pure stearyl alcohol to mixed monolayers containing stearic acid and stearyl alcohol in the 9:1 molar ratio. This linear variation between the ratios 9:1 and 0:1 of stearic acid and stearyl alcohol suggests that the two unit structures at each end are simply changing the relative proportions in the mixed mono-layers. However, the average potential for pure stearic acid monolayer on $CaCl_2$ is considerably below the value obtained by the extrapolation of this experimental straight line. This reduction in average potential for stearic acid monolayer in the presence of $CaCl_2$ is presumably due to the strong copoly-meric structure in stearic acid monolayer. In other words, we propose that the copolymeric structure may exist in the pure stearic acid monolayers, but that upon addition of more than 10 mole % of stearyl alcohol, the mixed monolayer consists of calcium distearate and stearyl alcohol. We further propose that the calcium distearate should have the structure such that none of the four oxygens of distearate are free to inter-act with the adjacent stearyl alcohol molecules. This condi-tion is required for the linear variation of average potentials in the presence of $CaCl_2$. If the carboxyl groups were inter-acting with the hydroxyl groups of the stearyl alcohol, then one would expect a non-linear decrease in average potentials from the additivity rule as shown in the presence of NaCl (Figure 2). Our proposed structure of calcium distearate is shown in Figure 6(c) which schemetically shows that all four oxygens of the two carboxyl groups probably interact with a calcium ion making a unit structure with no interaction with adjacent stearyl alcohol molecules. It is also interesting to note from Table I that the solid state of the stearic acid monolayer on subsolutions containing $CaCl_2$ is converted into the gel state by the presence of 10 mole % of stearyl alcohol in the mixed monolayer. This observation supports the con-clusion that a minimum of about 10 mole % stearyl alcohol disrupts the copolymeric structure.

It is also evident from Figure 5 that the structural arrangement of stearic acid and stearyl alcohol in the 3:1 and 1:3 molar ratios show kinks in the maximum value of the surface potential at the collapse pressure. The hexagonal packing seems to be occuring both in the 3:1 and 1:3 molar ratios when the mixed monolayers are compressed up to their collapse pressure (Figure 5). However, we did not see the striking change in foam stability or retardation of evapora-tion at 3:1 molar ratio since the average area per molecule at the 3:1 molar ratio is much greater and hence, the inter-action between the components considerably weaker as compared to that in the 1:3 mixed monolayer.

In summary, the results presented in this paper indicate the association and a closer molecular packing of stearic acid and stearyl alcohol at the 1:3 molar ratio and a structural

Table I. The State of Various Monolayers near their Collapse
Pressures (≈40 dynes/cm)

Monolayers		pH 8.8	
Stearic acid	Stearyl alcohol	tris buffer + 0.02 M NaCl	tris buffer + 0.01 M CaCl$_2$
100	0	gel	solid
90	10	gel	gel
75	25	gel	gel
67	33	gel	gel
50	50	gel	gel
33	67	gel	gel
25	75	gel	gel
10	90	viscous liquid	viscous liquid
0	100	viscous liquid	viscous liquid

alteration of the mixed monolayers at the 9:1 molar ratio. At
pH 8.8, the stearic acid monolayer forms a copolymeric
structure in the presence of calcium ions. This monolayer is
in the solid state. If 10 or more mole % of stearyl alcohol is
added, the mixed monolayers are in the gel state presumably
due to disruption of the copolymeric structure. The proposed
structure of calcium distearate does not allow any ion-dipole
interaction with the neighbouring stearyl alcohol molecules.

Summary

     Surface pressures, surface potentials, and the changes in
the state of mixed monolayers of stearic acid and stearyl
alcohol were determined on tris buffer + NaCl or tris buffer +
$CaCl_2$ at pH 8.8. For subsolutions of tris buffer + NaCl, the
average area per molecule showed optimum reduction at 9:1 and
1:3 molar ratios of stearic acid to stearyl alcohol. The
condensation at the 9:1 ratio appeared to be due to structural
alterations in the mixed monolayer whereas that at the 1:3
molar ratio was shown to be due to a closer packing of molecules. In the presence of $CaCl_2$ in the subsolution, the average area per molecule followed the additivity rule of molecular areas.

     For subsolutions of tris buffer + NaCl, the average
potential per molecule showed a nonlinear deviation from the
additivity rule, presumably due to ion-dipole interaction
between the ionized stearic acid and stearyl alcohol molecules. However, in the presence of $CaCl_2$ in the subsolution,
the average potential per molecule showed a linear variation
which was above the line representing the additivity rule.
Two dimensional lattice structure of these mixed monolayers
as well as the ionic structure of calcium distearate are
proposed to account for the experimental results.

Literature Cited

1.   Ross, J., J. Phys. Chem., (1958), 62, 531.
2.   Shah, D. O., J. Colloid Interface Sci., (1970), 32, 570.
3.   Shah, D. O., J. Colloid Interface Sci., (1970), 32, 577.
4.   Goddard, E. D., and J. H. Schulman, J. Colloid Sci.,
     (1953), 8, 309.
5.   Goddard, E. D., and Schulman, J. H., J. Colloid Sci.,
     (1953), 8, 329.
6.   Barnes, G. T., and LaMer, V. K., in "Retardation of
     Evaporation by Monolayers," V.K. LaMer, Ed., (1962),
     Academic Press, New York, 9-33.
7.   Rosano, H., and LaMer, V. K., J. Phys. Chem., (1956),
     60, 348.
8.   Dervichian, D. G., in "Surface Phenomena in Chemistry
     and Biology," J.F. Danielli, Ed., K.G.A. Pankhurst,

and A.C. Riddiford, Eds., pp. 70-87, Pergamon Press,
New York, 1958.
9.   Shah, D. O., Dysleski, C. A., J. Am. Oil Chem. Soc.,
     (1969), $\underline{46}$, 645.
10. Shah, D. O., and Schulman, J. H., J. Colloid Interface
     Sci., (1967), $\underline{25}$, 107.
11. Shah, D. O., J. Colloid Interface Sci., (1971), $\underline{37}$,
     744.
12. Shah, D. O., and Schulman, H. J., J. Lipid Res.,
     (1967), $\underline{8}$, 215.
13. Marsden, J., and Schulman, J. H., Trans. Faraday Soc.,
     (1938), $\underline{34}$, 748.
14. Schulman, J. H., and Hughes, A. H., Biochem. J., (1935),
     $\underline{29}$, 1243.
15. Cockbain, E. G., and Schulman, H. J., Trans. Faraday Soc.,
     (1939), $\underline{35}$, 716.
16. Harkins, W. D., and R. T. Florence, J. Chem. Phys.,
     (1938), $\underline{6}$, 847.
17. Florence, R. T., and Harkins, W. D., J. Chem. Phys.,
     (1938), $\underline{6}$, 856.
18. Myers, R. J. and Harkins, W. D., J. Phys. Chem., (1936),
     $\underline{40}$, 959.
19. Sanders, J. V., and Spink, J. A., J. Colloid Sci., (1955),
     $\underline{175}$, 644.
20. Matsubara, A., Matuura, R., and Kimizuka, H., Bull. Chem.
     Soc. (Japan), (1965), $\underline{38}$, 369.
21. Kimizuka, H., Bull. Chem. Soc. (Japan), (1956), $\underline{29}$,
     123.
22. Goddard, E. D., and Ackilli, J. A., J. Colloid Sci.,
     (1963), $\underline{18}$, 585.
23. Goddard, E. D., Smith, S. R., and Kao, O., J. Colloid
     Sci., (1966), $\underline{21}$, 320.
24. Rogeness, G., and Abood, L. G., Arch. Biochem. Biophys.,
     (1964), $\underline{106}$, 483.
25. Sears, D. F., and Schulman, J. H., J. Phys. Chem., (1964),
     $\underline{68}$, 3529.
26. Wolstenholme, G. A., and Schulman, J. H., Trans. Faraday
     Soc., (1951), $\underline{47}$, 788.
27. Schulman, J. H., and Dogan, M. Z., Disc. Faraday Soc.,
     (1954), $\underline{16}$, 158.
28. Wolstenholme ᵕ. A., and Schulman, J. H., Trans. Faraday
     Soc., (1950), $\underline{46}$, 475.
29. Sasaki, T., and Muramatsu, M., Bull. Chem. Soc. (Japan),
     (1956), $\underline{29}$, 35.
30. Shah, D. O., and Schulman, J. H., J. Lipid Res., (1965),
     $\underline{6}$, 341.
31. Shah, D. O., and Schulman, J. H., Lipids, (1967), 2, 21.
32. Long, C., Ed. "Biochemists' Handbook," p. 30, Van Nostrand,
     Princeton, 1961.

33. Shah, D.O., and Schulman, J.H., in "Molecular Association
     in Bilogical and Related Systems," Adv. Chem. Ser. No. 84,
     p.189, American Chemical Society, Washington, D.C., 1968.
34. Deamer, D.W., and Cornwell, D.G., Biochem. Biophys. Acta,
     (1966), 116, 555.
35. Deamer, D.W., Meek, D.W., and Cornwell, D.G., J. Lipid
     Res., (1967), 8, 255.

Contact Angles in Newton-Black Soap Films Drawn from Solutions Containing Sodium Dodecyl Sulphate and Electrolyte

J. A. DE FEIJTER and A. VRIJ

van't Hoff Laboratory for Physical and Colloid Chemistry,
Padualaan 8, Utrecht, The Netherlands

Introduction

In this symposium in which we honor the Professors R.D. and M.J. Vold for their many and significant contributions to Colloid and Surface Science, we will call the attention to a subject positioned somewhere between these two disciplines, i.e. to soap films. Soap films, those very thin liquid lamellae as found in soap bubbles, can serve as model systems for studying forces that are important for the understanding of colloids in general. Moreover, in a more direct way, soap films are model systems for the thin, liquid lamellae as found in foams, emulsions and biological membranes.

Here we will review some recent results (1) of contact angle measurements in very thin, so-called Newton Black (NB) soap films having a thickness of about 5 nm. The full details will be published elsewhere (2).

Contact Angles

When a soap film is drawn in a frame from a soap (detergent) solution it often occurs that after sufficient thinning of the film by drainage a contact angle appears at the transition between film and bulk solution. This necessarily means that the surface tension of the film surface, σ^f, is less than the surface tension of the bulk surface, σ^α, and that both are interconnected by the relation,

$$\sigma^f = \sigma^\alpha \cos \theta \tag{1}$$

where θ is the contact angle. At equilibrium the difference between σ^f and σ^α can be ascribed to a lowering of the film free energy, ΔF_e, due to the proximity of the film surfaces (3-6)

$$\Delta F_e = 2\sigma^f - 2\sigma^\alpha \tag{2}$$

In many cases ΔF_e is caused by double layer repulsion and van der Waals attraction forces in the film. Thus contact angle measurements can be used for the study of these forces. From the Equations (1) and (2) it follows that

$$\Delta F_e = 2\sigma^{\alpha}(\cos \theta - 1) \tag{3}$$

In our study the contact angles were measured by the diffraction technique of Princen and Frankel (7).

Results. ΔF_e as obtained from θ, with the help of Equation (3), for films drawn from sodium dodecyl sulphate solutions as a function of the concentration of added NaCl at several temperatures, is given in Figure 1. The film thickness was obtained from the reflection coefficient of light and expressed as the thickness of a pure water layer (so-called "equivalent water thickness", h_w, (6). It was constant in all cases and equal to 4.4 nm. The (bulk) surface tension, σ^{α}, at 23°C varied between 33.1 → 31.1 mJm^{-2} for NaCl-concentrations ranging between 0.19 → 0.50 mol dm^{-3}. The variation of σ^{α} with temperature was small and of minor importance in the calculation of ΔF_e from Equation (3).

Discussion

Figure 1 shows some striking features. In the first place below a certain NaCl-concentration ΔF_e is very small as indicated by the drawn horizontal line. Above this concentration the contact angle, θ, and thus also ΔF_e increases steeply. Apparently, there is some kind of transition between two states. The film in the first state is called "common black" (CB) film, the film in the second state is called "Newton black" (NB) film (8). In the second place one observes that the transition is very dependent on temperature.

Already Huisman and Mysels (6) showed that the great variation of ΔF_e with NaCl-concentration (C_3(NaCl)) cannot be explained by the classical picture where ΔF_e is composed of diffuse double layer and van der Waals interactions. It will be clear that also the large temperature dependence of ΔF_e, which we have measured systematically for the first time, as far as we know, is totally unexplainable by the classical picture.

We, therefore, turned our attention to a coherent thermodynamic description of such films in order to investigate what kind of information could be extracted from the slopes of ΔF_e vs. C_3, which are rather independent of temperature.

Gibbs Adsorption Isotherm for Films. It can be shown (1, 2) that the Gibbs adsorption equation for a film containing n components, and which is in equilibrium with the bulk solution and the surrounding vapor phase, can be put in the following form,

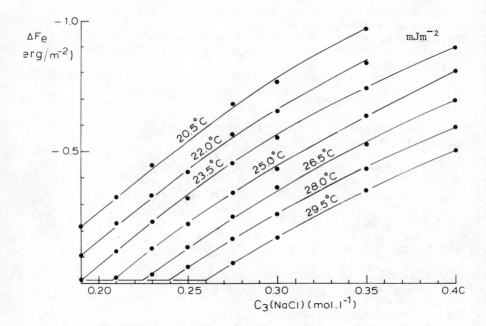

Figure 1.   Interaction free energies, ΔF_e, for films drawn from an aqueous solution containing 0.05% Na-dodecyl sulphate and NaCl at several temperatures.

$$d\gamma = -2 {}^s s^f dT + h d\Pi - \sum_{i=2}^{n} 2\Gamma_i^f d\mu_i \tag{4}$$

Here, γ is the film tension and Γ_i^f is the surface excess of component i in the film per unit surface area, chosen with respect to a Gibbs dividing surface in which the surface excess of the solvent, Γ_i^f (i=1 is solvent), is zero. The Gibbs dividing surfaces in the film are separated by a distance, h, from each other. The composition of the film is then completely defined with respect to the composition of the outer vapor phase and of an inner film reference phase (bulk phase) taken as continuous up to the dividing surfaces. Further, ${}^s s^f$ is the surface excess entropy per unit area, Π is the pressure difference between the vapor phase and the bulk phase (at the film border) which is called disjoining pressure, and T and μ_i are the temperature and the chemical potential of component i, respectively.

The film tension, γ, is connected with ΔF_e as follows (3–5),

$$\Delta F_e = - \int_{\infty}^{h_e} \Pi(h) dh \tag{5}$$

Upon integration of Equation (4) at constant T and μ_i (i $\geqslant$ 2) it then follows that

$$\gamma = 2\sigma^\alpha + \Delta F_e + \Pi h_e \tag{6}$$

The Gibbs adsorption equation for the bulk surface reads

$$d\sigma^\alpha = - {}^s s^\alpha dT - \sum_{i=2}^{n} \Gamma_i^\alpha d\mu_i \tag{7}$$

which substracted from Equation (4) gives

$$d(\Delta F_e + \Pi h_e) = - 2\Delta^s s dT + h d\Pi - 2 \sum_{i=2}^{n} \Delta\Gamma_i d\mu_i \tag{8}$$

where $\Delta^s s = {}^s s^f - {}^s s^\alpha$ and $\Delta\Gamma_i = \Gamma_i^f - \Gamma_i^\alpha$ are excess quantities of the film surface over that of the bulk surface. In our case i=1 is the solvent, i=2 is sodium dodecyl sulphate and i=3 is NaCl.

Surface Excess of Salt and Detergent. It will be apparent from Equation (8) that $\Delta\Gamma_2 = \Delta\Gamma_{NaDDS}$ and $\Delta\Gamma_3 = \Delta\Gamma_{salt}$ can be obtained from $\partial\Delta F/\partial\mu_2$ and $\partial\Delta F/\partial\mu_3$, respectively. From the graphs in Figure 1 and known mean activity coefficients $\Delta\Gamma_3$ can be calculated. (The term Πh is negligeable in our case and a slight correction for the change of μ_2 with μ_3 was taken into account.) The results are given in Table I.

Table I. Calculated Values of the Adsorption of Electrolyte in NB-soap-films drawn from an Aqueous Solution containing 0.05% Na-dodecyl Sulphate and NaCl respectively Na_2SO_4.

$\Delta\Gamma_3$ x 10^{11} (equivalent cm^{-2})						
T $^\circ$C \ salt	20.5	22.0	23.5	25.0	26.5	29.5
NaCl	1.39	1.35	1.34	1.33	1.29	1.22
Na_2SO_4		1.72	1.71			

$\Delta\Gamma_2$ was found to be small (< 0.5 x 10^{-11} mol cm^{-2}). At the bulk surface $\Gamma_2^\alpha = 42$ x 10^{-11} mol cm^{-2}. $\Delta\Gamma_3$ is the surface excess per unit area at a film surface minus that at the bulk surface. It will be clear from Table I that $\Delta\Gamma_3$ is positive and only slightly dependent on temperature. Further, $\Delta\Gamma_3$ is for Na_2SO_4 larger than for NaCl.

We first investigated what the diffuse electrical double layer predicts about $\Delta\Gamma_3$. At single electrical double layers counterions are attracted and co-ions are expelled from the surface. The overall effect is an expulsion of the (neutral) electrolyte (so-called negative adsorption). For high surface potentials $\Gamma_3^\alpha \simeq -2c_3\kappa^{-1}$, where κ^{-1} is the diffuse double layer thickness. In a film the electrical double layers overlap and Γ_3^f will be smaller in magnitude, i.e. $\Delta\Gamma_3$ is positive in accordance with the experimental results in Table I. The precise calculation of $\Delta\Gamma_3$ for diffuse, overlapping double layers is complicated and involves elliptical integrals (1, 2).

We made an attempt to interpret the numbers in Table I (at 23.5°C) in the following way. In our system the diffuse part of the double layers is probably small. Moreover in the film we expect a nearly total overlap of these diffuse double layers, so that $\Gamma_3^f \sim 0$ and $\Delta\Gamma_3 \simeq 0 - (-\Gamma_3^\alpha) \simeq \Gamma_3^\alpha$. Thus this would mean that we can (in an indirect way) obtain information about the diffuse double layer at the single (bulk) surface. The distance between the Outer Helmholz (OH)-planes in the film, 2d, is difficult to estimate. From the equivalent water thickness, h_w, of the film and an assessed density and refractive index of the hydrocarbon chains we obtained a thickness $h_2 = 1.6$ nm for the aqueous core of the film. Substracting from $\frac{1}{2}h_2$ the diameter of the hydrated SO_4'-group (say, 0.48 nm) and the radius of the co-ion (Cl') (say, 0.24 nm) gives for d a value in the order of one or a few tenths of nm. For our further calculations we took two choices, i.e. d = 0.1 nm and d = 0.4 nm. From the $\Delta\Gamma_3$ value of Table I, the potential, ψ_0^∞, at the OH-plane of the bulk surface was calculated, either by assuming that the potential,

ψ_0, at the OH-planes in the film was equal to ψ_0^{∞} or equal to $-\infty$. The results are given in Table II.

Table II. Calculated ψ_0^{∞}-values of the Bulk Surface, for the System NaDS + NaCl at 23.5°C. The Last Column gives the ζ-potentials of the NaDS-micelles.

C_3 (NaCl)	d = 0.1 nm		d = 0.4 nm		
(mol dm^{-3})	$\psi_0 = -\infty$	$\psi_0 = \psi_0^{\infty}$	$\psi_0 = -\infty$	$\psi_0 = \psi_0^{\infty}$	ζ
0.2	-41mV	-40mV	-75mV	-70mV	-73mV
0.3	-34	-32	-70	-64	-70
0.4	-30	-28	-70	-63	-68
0.5	-28	-25	-71	-62	-66

It will be apparent from this table that the choice of ψ_0 is not critical which indeed means that there is nearly total overlapping of double layers. The obtained values for ψ_0^{∞} seem reasonable in magnitude but are still very sensitive to the choice of d. The values for d = 0.4 nm conform rather well with ζ-potentials calculated from micelle mobilities (9). Similar calculations were performed for a(1-2) electrolyte (Na$_2$SO$_4$). These predict a ratio of 1.42 between the $\Delta\Gamma_3$-values of the two systems; the experimental value for this ratio from Table I is found to be 1.28.

Excess Film Entropy. From Equation (8) it follows that Δ^Ss can be derived from the variation of ΔF_e, and thus of θ, with temperature. In practice, however, Δ^Ss is not a very convenient quantity, as it is quite difficult to keep the μ_i's constant when T is varied. Instead of Δ^Ss we shall therefore use the interaction entropy, ΔS_e, defined by

$$\Delta S_e = -d\Delta F_e / dT \tag{9}$$

From Figure 1, ΔF_e vs T-plots can be obtained which are almost straight lines with slopes that are independent of C_{NaCl}. From this it follows that $\Delta S \simeq - 0.066 \pm 0.003$ mJm^{-2} K^{-1}, in all cases. For the CB-films the absolute value of ΔS is at least a factor of 50 smaller.

From ΔF_e and ΔS_e one can find the interaction energy, ΔU_e in the film

$$\Delta U_e = \Delta F_e + T\Delta S_e \tag{10}$$

Because ΔF_e varies from -0.01 to -1 mJm^{-2}, the value of ΔU_e is approximately -20 mJm^{-2}. Thus the formation of an NB-film from a CB-film is energetically a highly favorable but entropically a highly unfavorable process, with the quantities ΔU_e and $-T\Delta S_e$ tending to compensate each other. The large (negative) value of ΔS_e implies that the NB-films have a rather organized structure, different from that in CB-films. This was also found by Jones and Mysels (10) and by den Engelsen and Frens (11).

It is unlikely that the large (negative) value of ΔU_e can be due to the van der Waals forces. We surmise that it is caused by specific electrostatic interactions between ions of different charge signs juxtaposed with respect to each other. No model is however available to describe this in quantitative terms at present.

Summary

Contact angles were measured with the light diffraction technique of Princen and Frankel. The dependence of contact angles on electrolyte concentration was interpreted in terms of adsorption differences of electrolyte between film and bulk surfaces. The contact angles were very temperature dependent which directs to a kind of "phase transition" between the Newton Black and the common black films as found previously. The large entropies and energies of interaction imply that the Newton Black film is an organized structured with specific interactions other than diffuse double layer and van der Waals forces.

Literature Cited

1. de Feijter, J.A., "Contact Angles in Soap Films", Ph.D. thesis, University of Utrecht, 1973.

2. J. Colloid Interface Sci., to be published.

3. Princen, H.M., "Shape of Fluid Drops at Fluid-Liquid Interfaces and Permeability of Soap Films to Gases", Ph.D. thesis, University of Utrecht, 1965.

4. Princen, H.M., J. Colloid Sci., (1965) 20, 156.

5. Derjaguin, B.V., Acta Physico Chim., (1940) 12, 181.

6. Huisman, F. and Mysels, K.J., J. Phys. Chem., (1969) 73, 489.

7. Princen, H.M. and Frankel, S., J. Colloid Interface Sci., (1971) 35, 386.

8. Everett, D.H., Pure Appl. Chem., (1972) 31 (4), 614.

9. Stigter, D. and Mysels, K.J., J. Phys. Chem., (1955) 59, 45.

10. Jones, M.N., Mysels, K.J. and Scholten, P.C., Trans. Faraday Soc., (1966) 62, 1336.

11. den Engelsen, D. and Frens, G., J. Chem. Soc. Faraday Trans. I, (1974) 70, 193.

Stratification in Free Liquid Films

J. W. KEUSKAMP and J. LYKLEMA

Laboratory for Physical and Colloid Chemistry of the Agricultural University, De Dreijen 6, Wageningen, The Netherlands

Introduction

Stratification is the occurrence of layer-like structures. In concentrated surfactant solutions it is a familiar phenomenon. The smectic liquid crystalline phase is a characteristic example. It is less known that under the proper conditions stratification occurs also in free surfactant films. Long ago this has been established by Johonnott (1) and Perrin (2) for aqueous films in air, drawn from concentrated Na-oleate solutions. Later, Bruil and Lyklema (3) observed the same for free aqueous films from concentrated solutions of Na- or Li-dodecylsulphate (NaDS or LiDS respectively). They inferred the occurrence of stratification from some peculiarities in the drainage behavior.

The normal drainage behavior, as observed from the intensity of reflected monochromatic light, is a continuous thinning till the first or common black film is formed, in some cases followed by a step-wise formation of the second black or Newton film. The thickness of the common black is sensitive to the electrolyte concentration. It can vary between 10 and 80 nm and is mainly determined by van der Waals compression and double layer repulsion. On the other hand, Newton films are essentially bilayers. Their thicknesses are not very sensitive to the electrolyte concentration but they form only if special electrolytes are present in sufficient amounts. For example, Na^+-ions promote but Li^+-ions inhibit the formation of Newton films of Na-dodecylsulphate films.

The nature of the transition between the common and Newton film is still not clear, but there are many indications that specific structural effects, induced by ions, play a role (4). See also (5, 6) for such inference from contact angle measurements.

Under conditions conducive to stratification, the film initially drains gradually. However, once a thickness of the or-

For paper 1 in this series, see ref. (3).
Detailed information obtainable on request from first author.

der of a few tens of nm has been reached (the precise value depending on the composition of the film) the further thinning pro-
ceeds step-wise. Several steps after each other can be observed,
denoted as the .... third, second, first order, the highest order
pertaining to the thickest film. Bruil and Lyklema (3) observed
that the thickness difference between consecutive steps was
larger than, but nevertheless comparable with the thickness of
the Newton films. This suggests that stratified films are built
up of repeating units of pseudo-Newton films, viz. aqueous slabs
flanked by a surfactant layer. Hence, the study of stratification
can be informative with respect to the factors determining the
Newton film formation.

The study to be described is an extension of (3). The main
emphasis is on the effects of the nature and concentration of
added electrolytes.

Materials and Methods

Surfactants. NaDS, LiDS and NH_4DS have been prepared following Dreger et al. (7). These chemicals contained small amounts of
dodecanol (DOH) as judged from shallow minima in the γ-log c
curves of 2.4, 3.3 and 2.4 mN m^{-1} respectively. However, it was
verified that DOH does not affect stratification even in concentrations up to 5% of the surfactant. The CMC's, as determined
conductometrically are 8.3×10^{-3}, 8.8×10^{-3} and 7.1×10^{-3} mol dm^{-3}
respectively. The first value agrees well with general experience
(8). CMC's of the other two compounds are not available in literature as far as we are aware. Comparison between these values
suggests that NH_4^+ binds stronger to the DS$^-$-anion, whereas Li$^+$
binds weaker than the Na$^+$-ion.

Electrolytes, obtained from Merck were of analytical grade
and used without further purification.

Thickness Measurements have been made on vertical films with
an area of a few cm^2 using a light reflection method essentially
identical to that used by Lyklema et al. (9). The measuring site
in the film was 5 mm above the meniscus. The wavelength of the
monochromatic light was 546 nm and the temperature was kept at
25.00 ± 0.01°C. In order to derive the real thickness d from the
equivalent water thickness d_w the procedure by Frankel and
Mysels (10) was adopted to correct for the differing optical
properties of the surfactant surface layers. For a three-layer
film, consisting of an aqueous core, flanked by two dodecyl-
sulphate layers of thicknesses 1.06 nm and index of refraction
1.43 the correction (d - d_w) is -0.4 nm. In order to arrive at
this figure, the index of refraction of the aqueous core (1.36)
was assumed to be identical to that of a 2 M solution of $NaHSO_4$,
NH_4HSO_4 or $LiHSO_4$. For a film with p of such units repeating,
we applied a correction of -0.4 p nm. Here, p is the order of the
stratification.

In order to prevent evaporation, the vapor pressure in the box has been carefully adjusted and always sufficient time was allowed for acclimatisation. The reported thicknesses are all the average of several measurements and are independent of details in the actual drainage pattern.

Results and Discussion

Drainage Behavior. Plates 1 and 2 are photographs of stratifying NaDS-films in an advanced state of drainage. The bright silver region is the $\lambda/4$ maximum, with d_w above 100 nm. Above this silver band several regions of differing thickness can be observed, distinguishable as different shades of grey. Their thickness can be properly measured if their area is large enough, which is only the case for the lower (i.e. blacker) orders. Up to five orders have been thus estimated, but the plates suggest the presence of more of them.

It is important to realize that the pth order develops often spontaneously in the middle of the $(p+1)^{th}$ order. This shows that the drainage is not a "gliding off" of one repeating unit from the remainder (or from between the remainder) but due to a spontaneous disproportionation. The fluid expelled by this process flows downward. In some cases it is visible as a brighter rim around a dark spot, in other cases as bright spots.

From this observation we conclude that upon drainage the film passes through a number of metastable states, the Gibbs energy as a function of thickness being of the shape of Figure 1. The hump between the consecutive minima, e.g. the transition activation Gibbs energy increases with decreasing order, as judged from the increasing stability of the lower orders.

The trend is (see below) that less orders can be observed and that the thicknesses of the minima are greater if the surfactant concentration is lower. The dashed line in Figure 1 gives the probable shape of $G(d)$ for a lower surfactant concentration.

Stratification occurs at surfactant concentrations lower than those at which in bulk not yet elongated flat or cylindrical micelles develop (11). In view of this, we conclude from the fact that laminar structures do develop as soon as a certain critical thickness is reached upon drainage, that this stratification is induced by the surfactant layers on the surfaces, and more so, the closer these surface layers are (i.e. the thinner the film). Consequently, for a thermodynamic description of the stratification, the mutual interaction between the layers must be taken into account, this in contradistinction to the micellization in bulk which can be adequately treated as a monomer-micelle equilibrium without accounting for the presence of other micelles.

Plate 1. Stratified film of a NaDS solution in a vertical rectangular glass frame

Plate 2. Stratified film of a NaDS solution in a different stage of drainage than the film of Plate 1

Effect of Nature and Concentration of Surfactant. Figure 2
collects data for films of LiDS, NaDS and NH_4DS in the absence of
electrolyte. The general trends agree with those of Ref. (3), although there are some minor discrepancies in that in our case the
thickness for NaDS is a few percent lower and that now small but
systematic differences between the surfactants with different
cation are observed: d (NH_4DS) > d (LiDS) > d (NaDS). Also
common (non-stratified) films are slightly thicker with Li^+ than
with Na^+ (12). This can be explained by the lesser specific adsorption of Li^+ on the surfactant surface with the ensuing
stronger electrostatic repulsion. This explanation agrees with
the observed CMC-order of LiDS and NaDS, but not for NH_4DS (see
above).

The influence of the surfactant concentration (c_{soap}) is two-
fold. In the first place at higher concentration more orders are
observable.This is probably attributable to the better promotion of
cooperative structure formation. The second effect, the decrease
of d with increasing c_{soap} is more difficult to explain. Bruil
and Lyklema (3) found a dependence on $c_{soap}^{1/3}$ for NaDS-films. We
found the same dependence for NH_4DS.

The various orders are roughly equidistant. The differences
$d_p - d_{p-1}$ = Δd decrease with increasing c_{soap}. At the highest
c_{soap} studied (0.30 M for LiDS, 0.60 for NaDS and 0.50 for NH_4DS)
they are 6.6, 5.5 and 6.3 nm respectively. The mutual differences
are not significant. These increments decrease further with c_{soap}
and upon extrapolation $c_{soap} \rightarrow \infty$ attain limiting values of
5.2 ± 0.6 nm. This tends to the thickness of a Newton film. Under
these conditions the stratified film is apparently built as repeating Newton-like layers.

Effect of Electrolytes. The influence of added electrolytes
(concentration c_s) can be summarized into two points:

(i) For a given order, d decreases somewhat with c_s. For
example, in a film drawn from a 0.24 M LiDS solution d_1 decreases
from 8.8 to 7.8 nm and d_2 from 15.9 to 15.5 nm if LiCl is added
up to 0.12 M. In a film, drawn from 0.24 M NH_4DS, d_1 decreases
from 9.3 to 8.0 nm if NH_4Cl is added up to 0.3 M. The decrease is
roughly linear. The measurements are not accurate enough to say
whether the steepness of this fall-off is different for the different orders. A possible explanation is sought in terms of
electrostatic shielding. Note that the concentrations are so high
that the Debye-Hückel limiting law, which would predict a $c_s^{\frac{1}{2}}$ dependency, is no longer valid.

(ii) Electrolytes inhibit stratification, but the inhibiting
effect is strongly dependent on the nature of the cation. Considering the occurrence of the second order above the first, it
appears that only 0.008 M of added NaCl suffices to subdue its
occurrence in NaDS-films, whereas as much as 0.14 M LiCl is needed to suppress the occurrence of more than one layer in LiDS-
stabilized films. For the NH_4^+-case this figure is 0.16 M.

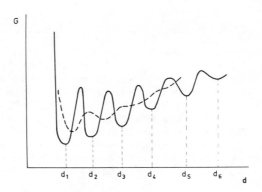

Figure 1. *Probable thickness dependency of the Gibbs energy of stratifying films. d_1, d_2 . . . etc. represent the thicknesses of the 1st, 2nd . . . order.* ———— *high surfactant concentration,* - - - - - - *low surfactant concentration.*

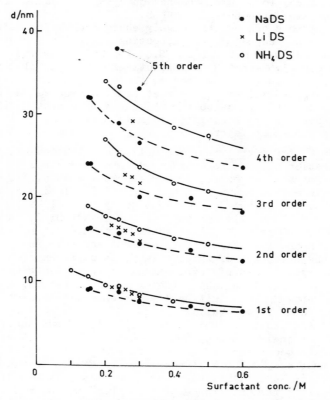

Figure 2. *Metastable equilibrium thicknesses of stratified films, stabilized by LiDS, NaDS, or NH_4DS. No electrolyte added,* $T = 25°C$.

There is a certain parallelism with the influence of electrolytes on common (not-stratified) black films: salts decrease
d, and Li^+- and NH_4^+-films are slightly thicker than Na^+-films,
but it depends critically on the nature of the cation whether or
not a jumpwise transition to the Newton film occurs. The difference is that in this case Na^+ promotes Newton film formation,
whereas Na^+ inhibits stratification. The nature of this contradiction is not clear, but phenomenologically it can be interpreted as a blunting of the activation energy humps in Figure 1,
and more strongly so in Na^+ than in Li^+, so that for common films
with Li^+ still one high barrier remains, preventing the Newton
film formation. The behavior of NH_4^+ compares better with that of
Li^+ than that of Na^+.

   The dramatic differences between Li^+ and Na^+ point out to a
very specific ionic effect. It is not known to us whether these
occur also in concentrated surfactant solutions. At any rate,
these phenomena deserve further attention.

Summary

Stratification of free liquid films is the occurrence in it
of layer-like structures. It can be observed if free films, drawn
from concentrated surfactant solutions, are drained. In such
films after some time several regions can then be seen, distinguishable in reflected light as different shades of gray. These
regions are denoted as orders, the lowest order being the thin-
nest film.

   The thicknesses of the various orders have been optically
determined for films drawn from NaDS, LiDS or NH_4DS, both in the
absence and presence of added electrolytes. The orders differ
between each other by an increment Δd that is roughly constant at
given surfactant concentration. For extremely high surfactant
concentrations Δd (as well as the thickness of the first order
film) approach that of a Newton (also known as a Perrin or second
black) film.

   Electrolytes inhibit the stratification, but Na^+-ions do so
far more effectively than Li^+- or NH_4^+-ions.

Acknowledgement

The authors acknowledge the skillful technical assistance of
Mr. R.A.J. Wegh.

Literature Cited

1. Johonnott, E.S., Phil. Mag. (6) (1906) 11, 746
2. Perrin, J., Ann. Phys. (1918) 10, 160
3. Bruil, H.G. and Lyklema, J., Nature Phys. Sci. (1971) 233, 19
4. Jones, M.N., Mysels, K.J. and Scholten, P.C., Trans. Faraday
   Soc. (1966), 62, 1336
5. Ingram, B.T., J. Chem. Soc. Faraday Trans. I (1972) 68, 2230

6. De Feijter, J.A., Ph.D. Thesis, State Univ. Utrecht (1973)

7. Dreger, E.E., Keim, G.I., Miles, G.D., Shedlovsky, L. and Ross, J., Ind. Eng. Chem. (1944) $\underline{36}$, 610

8. Mukerjee, P. and Mysels, K.J. "Critical Micelle Concentrations of Aqueous Surfactant Systems," NSRDS-NBS 36, p.51 ₁ Superintendent of Documents, Washington, D.C., 1971

9. Lyklema, J., Scholten, P.C. and Mysels, K.J., J. Phys. Chem. (1965) $\underline{69}$, 116

10. Frankel, S.P. and Mysels, K.J., J. Appl. Phys. (1966) $\underline{37}$, 3725

11. Reiss-Husson, F. and Luzzati, V., J. Phys. Chem. (1964) $\underline{68}$, 3504

12. Lyklema, J., Recl. Trav. Chim. Pays-Bas (1962) $\underline{81}$, 890

Surface Chemical Properties of Highly Fluorinated Polymers

MARIANNE K. BERNETT and W. A. ZISMAN

Laboratory for Chemical Physics, Naval Research Laboratory, Washington, D.C. 20375

Introduction

Highly fluorinated linear polymers are characterized by a low free surface energy and concomitant low wettability, as evidenced by the large contact angles of drops of organic and aqueous liquids. A comprehensive set of workable principles had been built up by Zisman and co-workers relating the chemical and spatial constitution in the outermost surface of a polymer with its surface energy (1, 2). During the last few years Wall, Brown, and Lowry of the National Bureau of Standards synthesized several new highly fluorinated ethylene polymers and copolymers (3-8) and established (4) that according to Wunderlich's "bead" theory and "rule of constant heat increment" (9, 10), the ethylenic polymers with fluorinated side groups generally contribute two carbon atoms to the backbone chain. This 2-carbon moiety plus the substituent constitute the smallest unit whose oscillations affect surface lattice equilibrium in a polymeric material, which represents the lowest free energy configuration.

This paper discusses the critical surface tensions of several of these radiation-induced polymers and copolymers, and compares them to those of polymers with related structures and surface constitutions.

Critical Surface Tensions of Wetting

Table I lists the experimental fluoropolymers and the data on their structural formulae, intrinsic viscosities [η] in hexafluorobenzene, and glass transition temperatures Tg (3-8), along with the code numbers used for this report. By casting a polymer from solution in hexafluorobenzene as a film on a clean glass slide, a smooth surface is obtained which is characterizable for wetting properties by contact angle (θ) measurements with freshly percolated liquids. Table II shows the average values of the advancing contact angles ($\pm$ 1°) obtained from such liquids of two homologous series, the n-alkanes and the

Table I. Physical Properties of Ethylenic Fluoropolymers

Code	Polymer and Composition, mol%	$[\eta]$, dl/g[a]	Tg, °C	γ_c, dyn/cm
I	$[-CH_2-CH(CF_3)-]_n$	1.10	27	21.5
II	$[-CH_2-CH(CF_3)-]_{58} + [-CF_2-CF_2-]_{42}$	5.00	19	21.1
III	$[-CH_2-CH(CF_3)-]_{39} + [-CF_2-CF_2-]_{61}$	2.10	9	18.0
IV	$[-CH_2-CH(CF_2CF_3)-]_n$	0.33	41	16.3
V	$[-CH_2-CH(CF_2CF_2CF_3)-]_n$	0.63	58	15.5
VI	$[-CH_2-CH(CF_2CF_2CF_3)-]_{79} + [-CF_2-CF_2-]_{21}$	0.74	45	16.2
VII	$[-CH_2-CH(CF_2CF_2CF_3)-]_{52} + [-CF_2-CF_2-]_{48}$	1.63	29	16.5
VIII	$[-CH_2-CH(CF_2CF_2CF_3)-]_{25} + [-CF_2-CF_2-]_{75}$	1.30	21	16.5
IX	$[-CH_2-CF(CF_3)-]_n$	4.65[b]	49	18.8
X	$[-CHF-CH(CF_3)-]_n$	0.23	87	17.5
XI	$[-CF_2-CF(n-C_5F_{11})-]_n$	$\overline{M}_n \approx 25000$	235	14.1
XII	$[-CH_2-CH(C_6F_5)-]_n$	4.0	105	22.5
XIII	$[-CF_2-CF(C_6H_5)-]_n$	1.0[c]	202	25.4
XIV	$[-CF_2-CF(C_6F_5)-]_n$	0.27	194	17.8

[a] in hexafluorobenzene; [b] in acetone; [c] in benzene

Table II. Advancing Contact Angles (Deg) of Liquids on Fluoropolymers (25°C)

Liquids	γ_LV (dyn/cm)	I	II	III	IV	V	VI	VII	VIII	IX	X	XI	XII	XIII	XIV
						Code									
n-Alkanes															
Hexadecane	27.8	50	51	63	61	67	64	61	59	52	56	66	41	14	52
Tetradecane	26.7	45	42	56	58	61	59	55	53	49	53	64	37	10	50
Tridecane	26.0				55					45	49	60	31	<5	46
Dodecane	25.4														
Undecane	24.6														
Decane	23.9	36	35	50	50	55	54	48	48	39	44	55	22		32
Nonane	23.1	29	29	47	46	51	51	45	44	36	41	53	13		28
Octane	21.8				42					30	36	49	9		25
Dimethylsiloxanes															
DC200 3.0 cSt	19.4					43	44	37	37				<5	<5	
DC200 2.0 cSt	18.9					40	40	33	33						
DC200 1.5 cSt	18.1					36	36	28	29					<5	
12 (DMS)	19.7				41					19	30	51			40
10 (DMS)	19.5				40					16	29	50			38
9 (DMS)	19.3				39					14	28	49			36
8 (DMS)	19.1				38					10	26	48			34
7 (DMS)	19.0				37					7	24	47			33
6 (DMS)	18.6				33					6	22	45			29
5 (DMS)	18.2				32					5	18	43			27

dimethylsiloxanes, where the latter were either the Dow Corning 200 series or the well-characterized series containing from 6 to 12 dimethylsiloxane (DMS) units. (11, 12). When the $\cos \theta$ of each member of such a homologous series of liquids on a smooth, solid, clean, low-energy surface is plotted against the surface tension (γ_{LV}) for each of those liquids, a straight line results; the intercept at $\cos \theta = 1$ ($\theta = 0°$) is referred to as the critical surface tension of wetting (γ_c) for that particular surface (13). Figure 1 shows such a graph for polymer IV, $[-CH_2-CH(C_2F_5)-]_n$, and is representative of the graphs for the other polymers with the exception of the perfluorophenyl - substituted polymers XII and XIV. Here the straight line for the n-alkanes displays a marked discontinuity in the region of γ_{LV} = 24-25 dyn/cm, resulting in two values of γ_c, differing by 1 dyn/cm. Experimental phenomena suggest that the smaller molecules are capable of slipping into the interstices of the poorly adlineated structures of polymers XII and XIV, whereas the larger and bulkier molecules are retained on the surface, thus being more reliable representatives for the true value of γ_c. Values of γ_c thus obtained for each polymer film are given in Table I.

Wettability for low-energy surfaces, as defined by γ_c, is determined essentially by the nature and packing of the exposed surface atoms of the solid and is otherwise independent of the nature and arrangement of the underlying atoms and molecules (1, 2). The arrangement of the surface atoms, of course, must represent the lowest free energy configuration for a given set of restraining conditions such as the nature, size of the underlying atoms, length of the chain, etc. In particular, for polymeric materials it depends on definition of the smallest unit, which for ethylenic polymers consists of 2 carbons in the backbone (9, 10) plus the substituent. Table III lists γ_c values of fluorine-containing ethylenic homopolymers, where the formulae shown are the repeating units in the polymer structure, as they would appear if all constituents were present in the surface. The order of structures is arranged to show progressive substitutions of either a hydrogen or a fluorine atom in the backbone chain by either a fluorine atom or a perfluoro group.

Several observations can be made from inspection of values of γ_c: (a) When the ethylene chain is fully hydrogenated, replacement of one hydrogen by a -CF$_3$ group lowers γ_c approximately 10 dyn/cm. (Compare XV and I; XVI and IX, and XVI and X.) (b) When one carbon atom in the 2-carbon moiety of the ethylene chain is fully fluorinated, replacement of a hydrogen by a -CF$_3$ group on the other carbon atom lowers γ_c only about 5 dyn/cm. (Compare XVII and XIX.) (c) Nearly equal γ_c values are observed on several pairs of monomers of unlike molecular constitutions, such as I and XVII, IX and XVIII, and X and XIX. Inspection of Stuart-Briegleb molecular models shows that, in certain steric arrangements, the γ_c-determining packing of fluorine atoms at the surface of these pairs could be very similar, since the pendant

Table III. Effect on Critical Surface Tensions of Wetting
by Replacement of Hydrogen with Fluorine and/or Pendant Group

Code	Polymer	γ_c (dyn/cm)
XV	$[CH_2-CH_2-]_n$	31^a
I	$[CH_2-CH(CF_3)-]_n$	21.5
IV	$[-CH_2-CH(CF_2-CF_3)-]_n$	16.3
V	$[-CH_2-CH(CF_2-CF_2-CF_3)-]_n$	15.5
XVI	$[-CH_2-CFH-]_n$	28^b
IX	$[-CH_2-CF(CF_3)-]_n$	18.8
X	$[-CH(CF_3)-CFH-]_n$	17.5
XVII	$[-CF_2-CFH-]_n$	22^b
XVIII	$[-CF_2-CF_2-]_n$	18.5^c
XIX	$[-CF_2-CF(CF_3)-]_n$	17^d
XI	$[-CF_2-CF(nC_5F_{11})-]_n$	14.1
XX	$[-CH_2-CH(C_6H_5)-]_n$	$33\text{-}35^e$
XII	$[-CH_2-CH(C_6F_5)-]_n$	22.5
XIII	$[-CF_2-CF(C_6H_5)-]_n$	25.4
XIV	$[-CF_2-CF(C_6F_5)-]_n$	17.8

[a]Reference 14; [b]Reference 15; [c]Reference 13; [d]Reference 16;
[e]References 17, 18

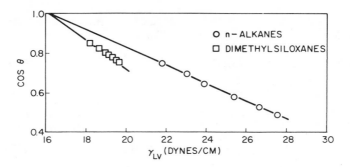

Macromolecules

Figure 1. Wettability of polymer $[—CH_2—CH(C_2F_5)—]_n$
(12)

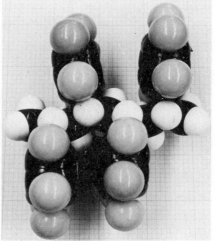

Macromolecules

Figure 2. Various configurations of polymer $[—CH_2—CH(C_6F_5)—]_n$. *(a) (top left) syndiotactic, exposure of flat side; (b) (top right) syndiotactic, exposure of edge; (c) (bottom left) isotactic* (12).

-CF$_3$ groups may be so located as to partially obscure the hydrogen atoms. The total effect achieves a balance between the hydrogen and the -CF$_3$ contributions which approximates the surface constitution of the linear unbranched configuration.
(d) Polymers IX and X demonstrate a confirmation of the conclusion of Pittman and co-workers (19), that in a particular fluorinated polymer γ_c is not necessarily dependent on the total fluorine content: despite the identical fluorine content, the molecular structures differ sufficiently to restrict free rotation of the -CF$_3$ groups in polymer X with the net result of closer surface packing of the fluorine atoms and thus a lower γ_c.
(e) As anticipated, the lowest γ_c, 14.1 dyn/cm, is obtained for polymer XI which is not only fully fluorinated but also displays the longest pendant perfluorinated group. The increase in length from 1 carbon to 5 carbons decreases γ_c about 3 dyn/cm (XIX and XI) because of better adlineation and less restriction in the longer chain. (f) For aromatic substitutions, replacement of the hydrogen with fluorine in either the phenyl group, XII, or the backbone chain, XIII, lowers γ_c about 10 dyn/cm from the 33-35 dyn/cm of polystyrene. Variations or spread of γ_c values for a given polymer can be now explained by the various orientations of the phenyl group in the surface, such as exposure of the flat side or the edge, or the tacticity of the arrangement (Figure 2 a, b, c). Total fluorination, as in polymer XIV of course results in even lower γ_c, since only fluorine atoms are exposed in the surface.

An interesting parallel can be observed and is demonstrated in Figure 3: When the hydrogen atoms in the ethylenic backbone are replaced by fluorine atoms, regardless of whether the pendant group is the alkyl -CF$_3$ or the aromatic -C$_6$F$_5$, γ_c is lowered by about 4.5 dyn/cm. When, on the other hand, the alkyl -CF$_3$ group is replaced by the aromatic -C$_6$F$_5$ group, whether on a fully hydrogenated or fully fluorinated backbone, γ_c is raised by approximately 1 dyn/cm.

Steric Configurations

The problem of determining the arrangement of the atoms and substituents of the first layer of the solid surface of a polymer or copolymer needs to be solved before we can relate the observed value of γ_c to the most probable surface composition of the polymeric solid. For simple unsubstituted polymers such as [-CF$_2$-CF$_2$-]$_n$ or [-CH$_2$-CH$_2$-]$_n$ the surface conformation can be rationalized by Stuart-Briegleb molecular models arranged in possible conformations on a flat table. However, where side chains are introduced into the model, steric hindrances are also introduced. It is obvious from Figure 4, where only some of the possible arrangements of the atoms and substituents in the first layer of the solid [-CH$_2$-CH(C$_2$F$_5$)-]$_n$ polymer surface are shown, that in a three-dimensional array a multitude of alternate

$$- CH_2 - CH(CF_3) - \qquad\qquad - CF_2 - CF(CF_3) -$$

21.5 $\xrightarrow{\;-4.5\;}$ 17.0

$\downarrow +1.0$ $+0.8 \downarrow$

$$- CH_2 - CH(C_6F_5) - \qquad\qquad - CF_2 - CF(C_6F_5) -$$

22.5 $\xrightarrow{\;-4.7\;}$ 17.8

Figure 3. Effect of fluorine or substituent replacement on
$\gamma_c \, (dyn/cm)$

Figure 4. Various configurations of polymer $[-CH_2-CH(C_2F_5)-]_n$. *(a) (top left) and (b) (top right) syndiotactic as viewed from two directions; (c) (bottom left) isotactic.*

conformations are possible, limited only by steric considerations. Lacking specific information on the tacticity of the radiation-induced polymer, we assumed them to be atactic, i.e., randomly oriented. Any prediction of surface packing is thus problematic. In addition to the spatial arrangements, polymers such as X also exhibit cis and trans isomerism with respect to the orientation of the -F and -CF_3 substituents on the backbone, further complicating and expanding the number of possible configurations.

Electrostatic Dipoles

At the juncture of the -$CH_2CH<$ backbone and the -$(CF_2)_x$F side group there exists an uncompensated electrostatic dipole. If one assumes that the side group is directed away from the polymer solid surface into the wetting liquid, the dipole is closer to the interface the lower the value of x. Experiments have shown that as x becomes smaller, θ also becomes smaller. Thus, when x changes from 1 to 3, γ_c is lowered from 21.5 to 15.5 dyn/cm (Figure 5), with the larger decrease of 5.2 dyn/cm when $1 < x < 2$ and the smaller decrease of 0.8 dyn/cm when $2 < x < 3$. In their study of adsorbed monolayers of progressively fluorinated fatty acids of the general formula $F(CF_2)_x(CH_2)_{16}COOH$ and $F(CF_2)_x(CH_2)_{10}COOH$, Shafrin and Zisman (20), showed that the uncompensated dipole has a large effect on wetting when $x \leq 7$, but becomes less significant when $x \geq 7$. Measurements of electrical and mechanical properties of insoluble monolayers on water by Bernett and Zisman (21), supported their view also. Shafrin and Zisman also noticed an abrupt reversal of the effect of homology at $2 < x < 3$ where the difference in γ_c was only 0.8 dyn/cm because of random tilting of the fluorocarbon group.

A similar abrupt discontinuity at $2 < x < 3$ for the ethylenic polymers can be explained by restriction of rotation of the substituent and the subsequent shielding of the electrostatic dipole. The latter is accentuated by the fact that the ethylenic hydrocarbon, to which the perfluoroalkyl side groups are connected, imposes restrictions on the packing of these substituent groups, necessitating progressively larger intramolecular rotations and bending within the chain. This results in exposure of the -CF_2- atomic grouping in an outermost surface of randomly oriented perfluoroethyl or -propyl groups. It would be interesting to study polymers with progressively longer perfluoroalkyl side groups to ascertain whether regular decreases in γ_c can be observed or whether a limiting value has been approached.

When the ethylene backbone is fully fluorinated, no large uncompensated dipoles are present, and a gradual and uninterrupted decrease in γ_c is observed with increase in x (Figure 5). The shape of the two curves in Figure 5 seems to point to an eventual asymptotic approach to the γ_c-vs-x curve obtained from adsorbed monolayers of fully fluorinated carboxylic acids, as shown in Figure 6 (20). Since to date no curve of any other family or series of related compounds (22), including the two present ones, has crossed or gone beyond the curve of the perfluoro acids, it is suggested that the latter represents an envelope of limiting values.

Copolymers

In their study on copolymers, Hu and Zisman (11) plotted γ_c for each polymer against the mole % of $[-CF_2-CF_2-]_n$ in the copolymer (Figure 7). The three graphs show the comparative case of predicting the effect on γ_c of copolymerization of $[CF_2-CF_2-]_n$ with $[-CH_2-CH_2-]_n$ (upper curve) (23) and the greater difficulty (middle graph) of predicting the effect on γ_c of copolymerization with $[-CH_2-CH(CF_3)-]_n$ and the intermediate effect with $[-CH_2-CH(C_3F_7)-]_n$ (bottom curve). It should be evident that we are dealing with several effects of adding fluorinated carbon atoms: (a) those due to London dispersion force changes (upper graph) and (b) those combining electrostatic effects and steric hindrance effects (lower two graphs).

It is obvious from the multitude of possible steric configurations and dipole contributions that we face the difficult problem of rationalizing how to compute the correct molecular conformation to arrive at a minimum surface energy for substituted ethylene fluoropolymers and copolymers.

Summary

Wetting properties of new, well-characterized, highly fluorinated linear ethylenic polymers and copolymers with tetrafluoroethylene were investigated. Fluorination in the polyethylene backbone was varied by degree of fluorine atom substitution; n-alkyl side chains of increasing number were fully fluorinated, whereas phenyl side groups were either fully or nonfluorinated. Critical surface tensions of wetting obtained on thin cast films of these fluoropolymers were compared to those of polymers with related structures and surface constitutions. Because of the presence of the bulky fluorine atoms and aromatic side groups, some of these molecules are extremely sterically blocked, which makes the prediction of an equilibrium surface conformation very difficult. The results are discussed in terms of solid surface constitution, steric hindrance, and electrostatic dipole contribution.

Literature Cited

1. Zisman, W. A., in "Contact Angle Wettability and Adhesion," Adv. Chem. Ser., No. 43, p. 1, Am. Chem. Soc., Wash., D.C., 1964.

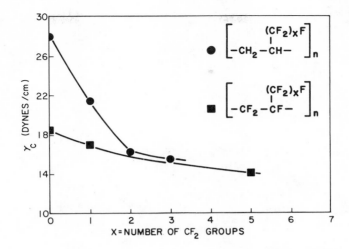

Figure 5.   Effect of substituent chain length on γ_c (12)

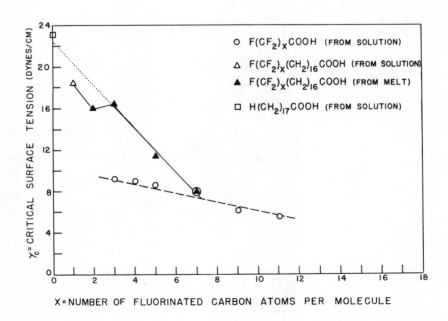

Figure 6.   Effect of fluorination of the adsorbed acid monolayer on γ_c (20)

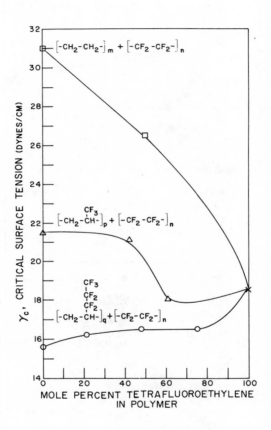

Figure 7. Effect of copolymerization with
$[—CF_2—CF_2—]_n$ *on* γ_c *(11)*

2. Zisman, W. A., J. Paint Tech., (1972), 44, 42.
3. Brown, D.W., Wall, L. A., J. Polym. Sci. A-1, (1968), 6, 1367.
4. Brown, D. W., Wall, L. A., J. Polym. Sci. A-2, (1969), 7, 601.
5. Brown, D. W., Lowry, R. E., Wall, L. A., J. Polym. Sci. A-1, (1970), 8, 2441.
6. Brown, D. W., Lowry, R. E., Wall, L. A., J. Polym. Sci. A-1, (1970), 8, 3483.
7. Brown, D. W., Lowry, R. E., Wall, L. A., J. Polym. Sci. A-1, (1971), 9, 1993.
8. Brown, D. W., Lowry, R. E., Wall, L. A., J. Polym. Sci. (Polym. Chem. Ed.) (1973), 11, 1973.
9. Wunderlich, B., J. Phys. Chem., (1960), 64, 1052.
10. Wunderlich, B., Bodily, D. M., Kaplan, H. M., J. Appl. Phys., (1964), 35, 95.
11. Hu, W. K. H., Zisman, W. A., Macromolecules, (1971), 4, 688.
12. Bernett, M. K., Macromolecules (In Press) (Sep 1974), 7.
13. Fox, H. W., Zisman, W. A., J. Colloid Sci., (1950), 5, 514.
14. Fox, H. W., Zisman, W. A., J. Colloid Sci., (1952), 7, 428.
15. Ellison, A. H., Zisman, W. A., J. Phys. Chem., (1954), 58, 260.
16. Bernett, M. K., Zisman, W. A., J. Phys. Chem., (1961), 65, 2266.
17. Ellison, A. H., Zisman, W. A., J. Phys. Chem., (1954), 58, 503.
18. Fox, R. F., Jarvis, N. L., Zisman, W. A., in "Contact Angle Wettability and Adhesion," Adv. Chem. Ser., No. 43, p. 317, Am. Chem. Soc., Wash., D.C., 1964.
19. Pittman, A. G., Sharp, D. L., Ludwig, B. A., J. Polym. Sci. A-1, (1968), 6, 1729.
20. Shafrin, E. G., Zisman, W. A., J. Phys. Chem., (1962), 66, 740.
21. Bernett, M. K., Zisman, W. A., J. Phys. Chem., (1963), 67, 1534.
22. Pittman, A. G., "Fluoropolymers," p. 419, Wiley Interscience Press, N.Y., 1972.
23. Fox, H. W., Zisman, W. A., J. Colloid Sci., (1952), 7, 109.

15

Effect of Surface Oxygen Complexes on Surface Behavior of Carbons

BALWANT RAI PURI

Department of Chemistry, Panjab University, Chandigarh, 160014, India

Introduction

It is now well know that microcrystalline carbons contain appreciable amounts of combined oxygen which gives rise to stable carbon-oxygen surface complexes (1). The work reported from our laboratories (2,3) and from elsewhere (4,5) indicates that there are definite surface groups or complexes which evolve carbon dioxide and, similarly, there are distinct surface entities which evolve carbon monoxide on evacuation at increasing temperatures. The effect of combined oxygen on various surface properties, such as wettability (6), heats of immersion in various liquids (7-9), adsorbability of water and other polar vapours (10-12), selectivity in adsorption from binary mixtures (13,14) has been reported. However, the effect of the individual oxygen complexes such as acidic CO_2-complex (15), nonacidic CO_2-complex (16) and CO-complex (3) on some of the surface properties mentioned above has not received adequate attention. An attempt has been made in the present paper to spell out the effect of each complex on (i) selective adsorption from mixtures of methanol and benzene, and (ii) adsorption of (a) benzene vapour, (b) dry ammonia, and (c) phenol from dilute aqueous solutions.

Experimental

Mogul (a colour black), Spheron-6 (a channel black) and a charcoal prepared by the carbonisation of recrystallised cane sugar (17) were used as such as well as after (i) outgassing at different temperature upto 1000°C, and (ii) treatment with aqueous hydrogen peroxide or potassium persulphate (16) so as to get materials associated with different amounts of surface oxygen complexes. The amount of combined oxygen and the form of its disposition was obtained by evacuating 0.5 g portion in a resistance tube furnace, raising the temperature to 1200°C and analysing the gases evolved in the usual manner (17).

The base adsorption capacity, estimated by titrating against aqueous barium hadroxide (17), fixed the amount of acidic CO_2-complex. This when subtracted from the amount of carbon dioxide evolved on evacuation gave the amount of non-acidic CO_2-complex (16) which arises through the fixation of oxygen at unsaturated sites (17). The amount of CO-complex was obtained from the amount of carbon monoxide evolved on evacuation. The nitrogen surface areas and the three main types of surface oxygen complexes of the various samples are recorded in Table I.

Selective adsorption from methanol-benezene solutions was studied by mixing 0.25 g carbon with a known weight (5-6 g) of the solution in a small glass tube drawn out at one end. The latter was sealed, after cooling in a freezing mixture, to minimise evaporation. Several such tubes containing 0.25 g carbon mixed with solutions of different concentrations were placed in a shaker (12 rev/min) held in a thermostat maintained at $35 \pm 0.05^{\circ}C$ for 48 hours. The change in composition of the liquid was determined interferometrically. Adsorption isotherms of benzene as well as dry ammonia were determined at $35 \pm 0.05^{\circ}C$ by using McBain's torsion balance technique. Adsorption isotherms of phenol from aqueous solutions were determined by mixing 0.5 g portions with a known weight (~ 5 g) of solutions of various concentrations, shaking the suspensions in a revolving wheel held in a thermostat at $35 \pm 0.05^{\circ}C$ for 24 hours. The change in concentration of the solution was determined interferometrically. The experiments were conducted in the low concentration range upto 20 mmoles/litre.

Results and Discussion

The composite adsorption isotherms, reproduced in Figure 1 from a recent report from the author's laboratory (18), show how the preference of Mogul for adsorption from methanol-benzene mixtures of various concentrations is altered in the presence of various surface oxygen complexes. It is seen that 1000°-outgassed Mogul, which is essentially free of combined oxygen, shows strong preference for benzene at all concentrations, giving a typically U-shaped isotherm (marked A). The 700°- outgassed Mogul, which is free of CO_2-complex but retains over 2.5 per cent oxygen capable of evolving carbon monoxide (CO-complex), shows, surprisingly enough, even greater preference for benzene, the less polar component of the mixture, all along the concentration range. This is contrary to the view generally held (13,14) that combined oxygen imparts greater preference for the more polar component of the mixture. It appears that the presence of quinonic groups which form a part of the CO-complex (19) promotes preference for benzene due to possibility of interaction of electrons of benzene ring with the partial positive charge on the carbonyl carbon atom (20). The 400°-outgassed and the original samples of Mogul, which contain increasing amounts of CO_2 com-

Table I. Surface Area and Surface Oxygen Complexes of the
Carbons used in the Present Work.

Description of the sample	Surface area m^2/g	Acidic CO_2-complex moles/ 100 g	Non-acidic CO_2-complex mmoles/ 100g	CO-complex mmoles/ 100 g
Mogul				
Original	308	68.5	nil	275
Outgassed at 400°C	306	42	nil	269
Outgassed at 600°C	314	nil	nil	198
Outgassed at 700°C	325	nil	nil	163
Outgassed at 850°C	318	nil	nil	52
Outgassed at 1000°C	326	nil	nil	nil
Original, treated with aq. H_2O_2	310	145	nil	284
Original, treated with $K_2S_2O_8$	312	189	nil	315
Outgassed at 1000°, treated with aq. H_2O_2	328	nil	120	20
Spheron-6				
Original	110	15.5	nil	122.5
Outgassed at 600°C	112	2.5	nil	108
Outgassed at 850°C	109	nil	nil	51
Outgassed at 1000°C	116	nil	nil	nil
Original, treated with aq. H_2O_2	114	61	nil	135
Outgassed at 1000°, treated with aq. H_2O_2	106	nil	10	4
Sugar Charcoal				
Original	412	355	nil	531
Outgassed at 600°C	514	51	nil	491
Outgassed at 1000°C	502	nil	nil	nil
Original, treated with aq. H_2O_2	443	453	nil	560
Outgassed at 1000° treated with aq. H_2O_2	488	nil	165	20

plex, though nearly equal amounts of CO-complex (cf. Table I), show, on the other hand, preferential adsorption of methanol over appreciable range of concentration (cf. isotherms marked C and D). This remarkable change in the preference may be ascribed to the presence of acidic CO_2-complex. This view receives support from the fact that when the amount of the acidic complex is raised from 68.5 mmoles to 145 mmoles on treatment with aqueous hydrogen peroxide and 189 mmoles on treatment with potassium persulphate, the preference of the surface is shifted increasingly in favour of methanol (cf. isotherms E and F). The comparison of the isotherms B and F shows clearly how the presence of a large amount of acidic CO_2-complex has brought about almost complete reversal of the preference of the carbon surface.

It is also interesting to note that the 1000°-outgassed Mogul yields almost identical isotherm after fixation of about 4 per cent of oxygen, as non-acidic CO_2-complex on treatment with aqueous hydrogen peroxide, as before fixation of any such oxygen (cf. isotherms marked G and A). The presence of non-acidic complex, evidently, produces little or no effect on the preference of the surface.

Thus the view held by previous workers (13,14) that the entire combined oxygen imparts polarity to the surface and promotes preference for the more polar component of a binary mixture needs revision in the light of the observations discussed above.

Adsorption of Benzene Vapour. The above conclusions were checked by studing adsorption of benzene vapour directly on a number of carbon blacks (21). The sorption-desorption isotherms (35±0.05°C) on some of the samples of Mogul (Figure 2) almost superimpose showing almost complete reversibility of the process. The amount of sorption at each relative vapour pressure is seen to increase appreciably as Mogul is outgassed at increasing temperatures. The maximum effect is produced when the black is outgassed at 600°C. It appears that with the elimination of the polar CO_2-complex and the emergence of CO-complex as the only predominant surface oxygen complex, the sorption of benzene increases appreciably on account of reasons advanced in the previous paragraph. The outgassing of Mogul at 850°C lowers the amount of CO-complex considerably and there is significant fall in the sorption value at all relative vapour pressures. With the complete elimination of the complex at 1000°C, there is a further fall in the sorption of benzene.

Adsorption of Dry Ammonia. Adsorption of dry ammonia on microcrystalline carbons at different temperatures has been investigated by a number of workers amongst which mention may be made of the work of Anderson and Emmett (22), Voskresenskii (23), Holmes and Beebe (24), Studebaker (25) and Dupupet et al (26. There is a general agreement that adsorption is enhanced appreciably in the presence of combined oxygen. However, the exact

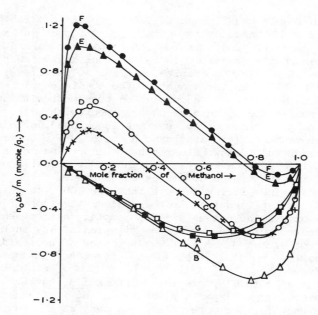

Figure 1. Composite adsorption isotherms of methanol–
benzene mixtures on Mogul before and after various treatments. A = Mogul outgassed at 1000°C, B = Mogul outgassed at 700°C, C = Mogul outgassed at 400°C, D =
Mogul original, E = Mogul original, treated with aq. H_2O_2,
F = Mogul original treated with acidified $K_2S_2O_8$, G =
Mogul outgassed at 1000°C, treated with aq. H_2O_2.

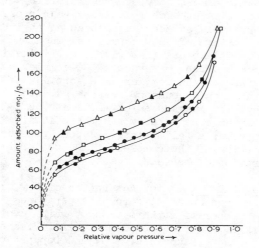

Figure 2. Adsorption isotherms of benzene on
Mogul. The solid points denote desorption
data. ○ = original, △ = outgassed at 600°C,
□ = outgassed at 850°C, ● = oxygen-free.

role of the various surface oxygen complexes, which constitute this oxygen, has not been elucidated.

Adsorption isotherms of ammonia on sugar charcoal, Mogul and Spheron-6 before and after various treatments are shown in Figures 3, 4 and 5, respectively. It is seen that, in each case, there is a considerable fall in the sorption of ammonia at each pressure on outgassing and that the effect is relatively more when a carbon is outgassed at 600°C, causing elimination of most of CO_2-complex (cf. Table I), than that when it is outgassed in 600-1000° temperature range, causing elimination of CO-complex. This shows that the sorption of ammonia is influenced more by the former than by the latter surface oxygen complex. The treatment of original samples with aqueous hydrogen peroxide which results in an appreciable rise in the amount of the acidic complex without affecting much the value of CO-complex (cf. Table I), is seen to cause increase in the sorption of ammonia at each vapour pressure, as can be seen on comparison of isotherms a and d in each of the Figures. The treatment of 1000° - outgassed carbons with aqueous hydrogen peroxide results in appreciable fixation of oxygen which, however, gives rise mostly to non-acidic CO_2-complex (cf. Table I). The isotherm on this sample is seen to be almost identical with that on the corresponding oxygen-free sample. These observations show clearly that the entire combined oxygen does not have a uniform effect on the sorption of dry ammonia by carbon. The oxygen present as acidic CO_2-complex influences the sorption of ammonia to a relatively larger extent than that present as CO-complex while the oxygen-present as non-acidic CO_2-complex has hardly any effect at all.

Adsorption isotherms on sugar charcoal, Mogul and Spheron-6, all previously outgassed at 1000° and therefore essentially free of oxygen, are plotted in Figure 6. It is seen that the various points fit around a single curve showing that the extent of sorption/m^2 of carbon surface when free of oxygen is about the same in every case irrespective of differences in porosity of these materials. It appears that ammonia, being a small molecule with thickness equal to 2.36 A, becomes easily accessible to the inner surface of carbons as well.

Adsorption of Phenol from Aqueous Solutions. Adsorption of phenol from aqueous solution, being of interest from the standpoint of water treatment, was studied using Mogul, Spheron-6, Graphon(a highly graphitised carbon black) and sugar charcoal. The isotherms (35° C) plotted on the basis of amounts adsorbed (μ moles)per m^2 of surface in the various carbons in the original state are presented in Figure 7. It is seen that the extent of adsorption at a given concentration is maximum in the case of Graphon which is free of oxygen and decreases in the order Graphon > Spheron-6 > Mogul > Sugar charcoal. This is also the order of decreasing oxygen content of these materials. The role of chemisorbed oxygen in adversely affecting the amount of adsorption is, therefore, quite evident.

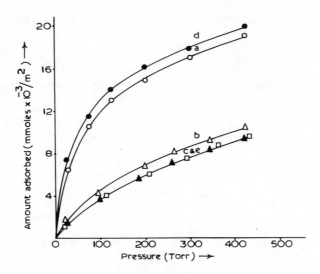

Figure 3.   Adsorption isotherms of ammonia on various samples of sugar charcoal. ◯ = original, △ = 600°-out-gassed, ☐ = 1000°-outgassed, ● = original, treated with aq. H_2O_2, ▲ = 1000°-outgassed, treated with aq. H_2O_2.

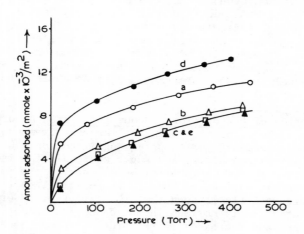

Figure 4.   Adsorption isotherms of ammonia on various samples of Mogul. ◯ = original, △ = 600°-outgassed, ☐ = 1000°-outgassed, ● = original, treated with aq. H_2O_2, ▲ = 1000°-outgassed, treated with aq. H_2O_2.

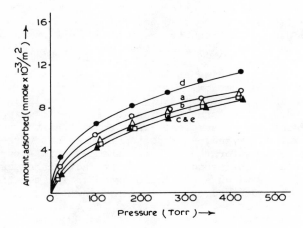

Figure 5.   *Adsorption isotherms of ammonia on various samples of Spheron-6.* ○ = *original,* △ = *600°-outgassed,* □ = *1000°-outgassed,* ● = *original, treated with aq.* H_2O_2, ▲ = *1000°-outgassed, treated with aq.* H_2O_2.

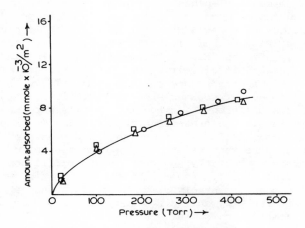

Figure 6.   *Adsorption isotherms of ammonia on oxygen-free (1000°-outgassed) samples of sugar charcoal, Mogul and Spheron-6.* ○ = *1000°-outgassed sugar charcoal,* △ = *1000°-outgassed Mogul,* □ = *1000°-outgassed Spheron-6.*

In order to study the effect of CO_2- and CO-complexes, separately, the isotherms were also determined on 600°- and 1000°-
outgassed samples. The results are plotted in Figures 8, 9 and
10 for Spheron-6, Mogul and Sugar Charcoal. It is highly significant to note that when the acidic CO_2- complex is eliminated
substantially on outgassing at 600° and CO-complex becomes the
predominant surface complex, the extent of adsorption at each
concentration increases appreciably in the case of each carbon
and even surpasses the corresponding values for the oxygen-free
(i.e., 1000°-outgassed) samples. This shows that the adverse
effect of combined oxygen, as reported by previous workers (27,
28), can not be true for the whole of the oxygen. A part of the
oxygen disposed of as carbon monoxide, in fact, enhances adsorbability of phenol to an appreciable extent. The positive effect
of a part of the combined oxygen in enhancing the sorption of
phenol is being reported, probably, for the first time. The results clearly indicate that when the acidic CO_2-complex is predominant, the surface prefers the strongly polar molecule of
water (the solvent), which adversely affects the adsorbability of
phenol. However, with the elimination of this complex and with
the emergence of CO-complex, as the predominant surface complex,
the preference of the surface for phenol rises appreciably. This
appears to be due to interaction of OH groups of phenol with
phenolic and quinonic oxygens associated with CO-complex.

Summary

The effect of carbon-oxygen surface complexes on selective
adsorption from methanol-benzene mixtures as well as adsorbability of benzene vapour, dry ammonia and phenol from dilute
aqueous solutions by a few samples of microcrystalline carbons
has been investigated. The view of previous workers that the
combined oxygen affects these properties more or less uniformly
has not been substantiated. Thus, while the presence of acidic
CO_2-complex enhances preference of the surface for methanol, the
more polar component of methanol-benzene solutions, that of CO-
complex enhances preference for benzene, the less polar component
of these solutions. The presence of non acidic CO_2-complex has
hardly any effect. Again, while the acidic CO_2-complex supresses
the sorption of pure benzene from vapour phase, that of CO-com-
plex enhances it appreciably and that of non acidic complex has
hardly any effect at all. Adsorption of dry ammonia, which for
oxygen-free carbons is largely a function of surface area, is
enhanced considerably by acidic CO_2-complex, to a much smaller
extent by CO-complex and not al all by non acidic CO_2-complex.
Adsorption of phenol from dilute aqueous solutions is influenced
adversely by acidic CO_2-complex, favourably by CO-complex but
not by non acidic complex. Suitable explanations have been
offered for the apparent anamolies.

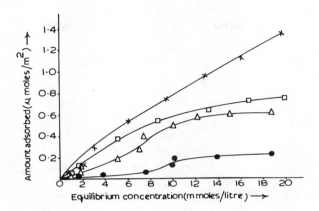

Figure 7. Adsorption isotherms of phenol on Graphon, sugar charcoal, Mogul, and Spheron-6. ✕ *= Graphon,* ● *= sugar charcoal,* △ *= Mogul,* □ *= Spheron-6.*

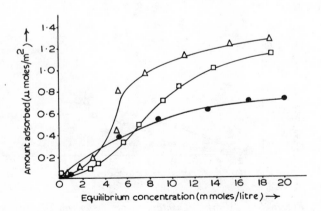

Figure 8. Adsorption isotherms of phenol on Spheron-6 before and after outgassing at 600 and 1000°C. ● *= original Spheron-6,* △ *= 600°-outgassed Spheron-6,* □ *= 1000°-outgassed Spheron-6.*

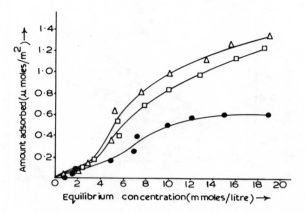

Figure 9. *Adsorption isotherms of phenol on Mogul before and after outgassing at 600 and 1000°C.* ● = *original Mogul,* △ = *600°-outgassed Mogul,* □ = *1000°-outgassed Mogul.*

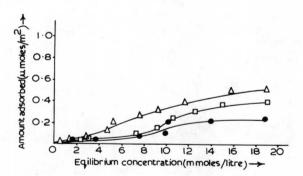

Figure 10. *Adsorption isotherms of phenol on sugar charcoal before and after outgassing at 600 and 1000°C.* ● = *original sugar charcoal,* △ = *600°-outgassed sugar charcoal,* □ = *1000°-outgassed sugar charcoal.*

Literature Cited

1. Puri, B.R., in "Chemistry and Physics of Carbon",
    P.L. Walker, Jr., Ed., Vol. VI, pp. 191-282, Marcel
    Dekker Inc., New York, 1970.
2. Puri, B.R., Myer, Y.P. and Sharma, L.R. Res. Bull.
    Panjab Univ., (1956) No. 88, 53.
3. Puri, B.R. and Bansal, R.C., Carbon, (1964) 1, 451.
4. Lobenstein, W.R. and Deitz, V.R., J. Phys. Chem.,
    (1955) 59, 481.
5. Rivin, D., Rubber Chem. Tech., (1971) 44, 307.
6. Studebaker, M.L. and Snow, C.W., J. Phys. Chem., (1955)
    59, 973.
7. Healey, F.H., Chessick, J.J., Zettlemoyer, A.C. and Young
    C.J., J. Phys. Chem., (1954) 58, 887.
8. Puri, B.R., Singh, D.D. and Sharma, L.R., J. Phys. Chem.,
    (1958) 62, 756.
9. Wade, W.H., J. Colloid Interface Sci., (1969) 31, 111.
10. Dubinin, M.M., Zaverina, E. D. and Serpinski, V.V.,
    J. Chem. Soc., (1955) 1760.
11. Kiselev, A.V. and Kovaleva, H.V., Izvest Akad. Nauk.
    SSSR, Otdel Khim. Nauk, (1959) 989.
12. Puri, B.R., Murari, K. and Singh, D.D., J. Phys. Chem.,
    (1961) 65, 37.
13. Gasser, C.G. and Kipling, J.J., J. Phys. Chem., (1960)
    64, 710.
14. Coltharp, M.T. and Hackerman, N., J. Colloid Interface Sci.,
    (1973) 43, 176; (1973) 43, 185.
15. Puri, B.R., Carbon, (1966) 4, 391.
16. Puri, B.R., Sharma, G.K. and Sharma, S.K., J. Indian Chem.
    Soc., (1967) 44, 64.
17. Puri, B.R., Sandle, N.K. and Mahajan, O.P., J. Chem. Soc.,
    (1963) 4880.
18. Puri, B.R., Singh, D.D. and Kaistha, B.C., Carbon, (1972)
    10, 481.
19. Studebaker, M.L., Huffman, E.W.D., Wolfe, A.C. and Nabors,
    L.G., Ind. Eng. Chem., (1956) 48, 162.
20. Bhacca, N.S., Tetrahedron Lett., (1964) 41, 3124.
21. Puri, B.R., Kaistha, B.C., Yasho Vardhan and Mahajan, O.P.,
    Carbon, (1973) 11, 329.
22. Anderson, R.B. and Emmett, P.H., J. Phys. Chem., (1952) 56,
    756.
23. Voskressenskii, A.A., Kolloid Zhur., (1961) 23, 3.
24. Holmes, J.M. and Beebe, R.A., J. Phys. Chem., (1957) 61,
    1684.
25. Studebaker, M.L., Rubber Age, (1957) 80, 661.
26. Dupupet, G., and Bastick, M., C.R. Acad. Sci. Ser., (1969)
    269, 437.

27. Clauss, A., Boehm, H.P. and Hoffman, U., Z. Anorg. Allgem. Chem., (1957) 290, 33.
28. Coughlin, R.W., Ezra, F.S. and Tan, R.N., J. Colloid Interface Sci., (1968) 28, 386.

Surface Properties of Nickel Hydroxide Before and After Dehydration to Nickel Oxide

M. TOPIC,* F. J. MICALE, C. L. CRONAN, H. LEIDHEISER, JR., and
A. C. ZETTLEMOYER

Center for Surface and Coatings Research, Lehigh University, Bethlehem, Penn. 18015

Introduction

It is a pleasure for us to take part in this symposium honoring Professors Marjorie and Robert Vold. They have done much for colloid science in teaching and in research through almost forty years of service; they have developed leaders in their field on several continents. As they come to well deserved retirement, we note their many contributions which will stand as a monument to their efforts.

The technique of thermal decomposition of precipitated $Ni(OH)_2$ for the preparation of NiO is often employed in studies (1-4) of this useful oxide. During surface characterization studies of $Ni(OH)_2$ and NiO a relationship has been found between the specific surface areas of the product and precursor. This relationship has allowed the proposal of a mechanism describing the process of thermal decomposition of $Ni(OH)_2$ at moderate temperatures.

Nickel hydroxide has the CdI_2 crystal structure (5) with the basic repeating layer structure. Each hexagonally packed layer of $Ni(OH)_2$ is made up of a hexagonal layer of Ni^{++} sandwiched between two hexagonal layers of OH^-. Nickel oxide has the NaCl structure.

Experimental

Nickel hydroxide samples were prepared using various techniques and are designated 1 through 4. Samples 1 and 2 were precipitated with NH_3 gas from solutions of $Ni(NO_3)_2$ at 80°C for sample 1 and at 25°C for sample 2. These samples were repeatedly washed with deionized distilled water and centrifuged until the wash water attained a constant conductivity of approximately 3×10^{-6} mho/cm. Samples 3 and 4 were obtained from Dr. Velimir Pravdic (Rudjer Boskovic Institute, Zagreb, Yugoslavia) and were

---

* Current address – Rudjer Boskovic Institute, Zagreb, Yugoslavia.

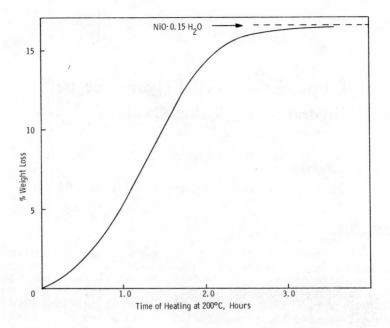

Figure 1.   Percent weight loss of Ni(OH)₂ vs. time at 200°C in vacuum

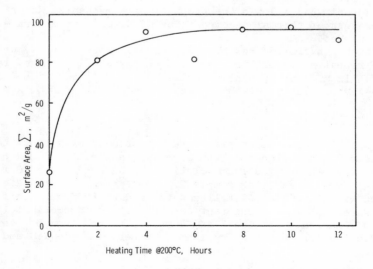

*Figure 2.   Specific surface area of sample vs. time of activation at
200°C in vacuum*

prepared by a modified sol-gel process (<u>6</u>). Specific surface area values were obtained using argon gas and a classical volumetric BET apparatus. Gas pressures were measured with a capacitive electronic manometer equipped with a 1000 torr differential pressure head (Datametrics, Inc.). All isotherms were determined with the samples immersed in a liquid nitrogen bath. Weight loss values were obtained through the use of a quartz spring microbalance (Worden) with a deflection sensitivity of approximately 1 mg/cm and sample capacity of about 30 mg.

Results and Discussion

Studies of the transition between $Ni(OH)_2$ and NiO involved the direct observations of weight loss and surface areas. Figure 1 illustrates a typical weight loss vs. time curve for a sample of $Ni(OH)_2$ at 200°C under vacuum conditions. The process includes a period of induction, passing through a maximum rate at about 1-1/2 hours and leveling off at 85% completion, corresponding to a stoichiometry $NiO \cdot 0.15H_2O$. This limit compares well with previously reported values of $NiO \cdot 0.16H_2O$ (<u>7</u>) and $NiO \cdot 0.18H_2O$ (<u>8</u>).

The surface areas of a sample of $Ni(OH)_2$ during the 200°C thermal treatment were measured at various activation times and are illustrated in Figure 2. As demonstrated in Figure 1, the decomposition is virtually complete after 3 hours of 200°C activation. On the basis of this work the thermal decomposition conditions for all samples were 200°C for 8 hours in vacuum.

The BET argon surface areas, assuming $13.8Å^2$ for the cross sectional area of argon, of the four $Ni(OH)_2$ samples and the resultant NiO samples prepared under the above conditions are given in Table I. Figure 3 is a plot of the ratio of the specific surface areas of NiO ot those of the corresponding precursor $Ni(OH)_2$ vs. the reciprocal of the specific surface area of the $Ni(OH)_2$ sample. The straight line drawn through the four sample points has a least squares slope and intercept of 73.9 m^2/g and 0.72, respectively.

The surface area relationship presented in Figure 3 and electron micrograph results, which show splitting $Ni(OH)_2$ hexagonal crystals into hexagonal layers, suggests a model of the thermal decomposition process. Figure 4 illustrates the proposed physical process for the decomposition. In the following derivation the subscript 1 refers to $Ni(OH)_2$ and 2 refers to NiO.

Electron micrographs showed that the precipitated $Ni(OH)_2$ occurs as hexagonal platelets which can be represented as having a radius R_1, and a thickness H_1 (greatly exaggerated in Figure 4). The process of 200°C activation results in the $Ni(OH)_2$ crystal exfoliating into <u>n</u> identical layers of NiO, each of radius R_2 and thickness H_2. The specific surface area, Σ, of a material consisting entirely of identical hexagonal plates and density ρ can be shown to be equal to:

Table I. Specific Surface Areas of
Ni(OH)$_2$ and NiO Prepared at 200°C

Sample	Ni(OH)$_2$	NiO
1	14	84
2	28	96
3	49	102
4	128	177

Table II. Physical and X-ray Properties

	Ni(OH)$_2$	NiO
M.W. (g/mole)	92.71	74.69
ρX-ray (g/cc)	3.950	6.809
$\overline{V}$ (cc/mole)	23.47	10.97
c_o	4.605$\overset{\circ}{A}$	a $= 4.1769$ $\overset{\circ}{A}$
a_o	3.126$\overset{\circ}{A}$	$d_{111} = 2.410$ $\overset{\circ}{A}$
		$d_{220} = 1.476$ $\overset{\circ}{A}$

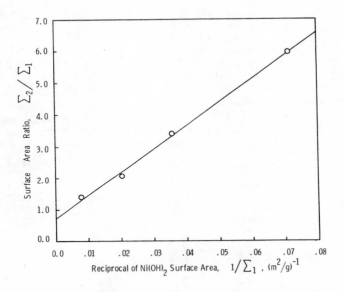

Figure 3. Ratio of NiO surface area to Ni(OH)₂ surface area vs. the reciprocal of the Ni(OH)₂ specific surface area

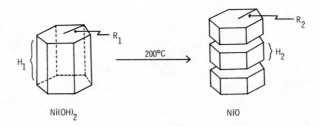

Figure 4. Schematic diagram of the exfoliation of Ni(OH)₂

$$\Sigma = \frac{2R + CH}{RH\rho} \tag{1}$$

where $C = \frac{4\sqrt{3}}{3} = 2.309,$

and R and H are the crystal radius and thickness. Table II contains various physical and X-ray properties (9) of Ni(OH)$_2$ and NiO. Most notable, the molar volume decreases by 53% during the decomposition process. Analysis of X-ray data and electron micrographs indicate that the broad hexagonal face of Ni(OH)$_2$ is the (001) plane. After thermal decomposition, the broad hexagonal face becomes the (111) plane of NiO. It can be shown that the following relationships exist between the crystal dimensions of Ni(OH)$_2$ and NiO:

$$H_2 = \frac{h}{n} H_1 \tag{2}$$

and $\quad R_2 = rR_1 \tag{3}$

where $\quad r = \frac{d_{110}}{a_o} = \frac{2.952}{3.126} = 0.9443 \tag{4}$

and $\quad h = \frac{d_{111}}{c_o} = \frac{2.410}{4.605} = 0.5233 \tag{5}$

Expressions for Ni(OH)$_2$ and NiO, using the form of Equation (1), can be combined if both Σ_1 and Σ_2 are written as functions of R_1 and H_2. Elimination of R_1 results in the equation:

$$\frac{\Sigma_2}{\Sigma_1} = \frac{2}{H_2\rho_2} \left(1 - \frac{h}{rn}\right) \frac{1}{\Sigma_1} + \frac{\rho_1}{r\rho_2} \tag{6}$$

where $\quad r = 0.9443$
$\quad\quad\quad\quad h = 0.5233$

Equation (6) represents a straight line with an intercept of $\rho_1/r\rho_2$ and slope of $(2/H_2\rho_2)(1 - [h/rn])$ if the surface areas are plotted as in Figure 3. Evidently the slope is a constant over the range of surface areas studied, even through it is a function of two possible variables, H_2 and n. Given that the experimental value of the slope in Figure 3 is 73.9 m^2/g and as shown in Equation (6):

$$S = \text{slope} = \frac{2}{H_2\rho_2} \left(1 - \frac{h}{rn}\right) \tag{7}$$

Replacing n by $\quad n = \dfrac{hH_1}{H_2}$

and rearranging $\quad H_2 = \dfrac{2rH_1}{S\rho_2 rH_1 + 2}$ $\qquad\qquad$ (8)

$$= \frac{1.889\ H_1}{0.04752\ H_1 + 2}$$

where H_1 and H_2 are expressed in Å units. Figure 5 is a plot of the thickness of an NiO crystal, H_2, vs. the thickness of its parent $Ni(OH)_2$ hexagonal plate, H_1, as calculated from Equation (8). The distance H_2 is the (111) directional thickness of the NiO hexagonal plate.

The experimental intercept of 0.72 agrees with the theoretical value of 0.614 calculated from Equation (6). The difference in the intercept value may be due to a real value of ρ_2 lower than the X-ray value presented in Table II, especially since the activation was only 85% complete. Other author's (2,10) have measured the NiO density as 6.0 g/cc. If this density applies to these samples the theoretical intercept should be 0.70, which is in excellent agreement with the experimental intercept.

Analysis of Equation (6) indicates that polydisperse $Ni(OH)_2$ samples yields a scatter of data if plotted as in Figure 3. Since the scatter is not severe in Figure 3 it can be concluded that either there is little polydispersity or the four samples are approximately equally polydispersed.

Literature Cited

1. Richardson, J.T., Milligan, W.O., Phys. Rev. (1956), 102, 1289.

2. Larkins, F.P., Fensham, P.J., Sanders, J.V., Trans. Faraday Soc., (1970), 66, 1748.

3. Nicolaon, G.A., Teichner, S.J., J. Colloid Interface Sci., (1972), 38, 172.

4. Fahim, R.B., Abu-Shady, A.I., J. Catal., (1970), 17, 10.

5. Huckel,W.,"Structural Chemistry of Inorganic Compounds", Vol. II, p. 547ff, Elsevier Publishing Co., Amsterdam, 1951.

6. Bonacci, N., Novak, D.M., Croat. Chem. Acta, (1973), 45, 531.

7. Teichner, J., Morrison, J.A., Trans. Faraday Soc., (1955), 51, 961.

8. Klier, K., Kinet. Katal. (Eng. trans.), (1962), 3, 51.

232

ADSORPTION AT INTERFACES

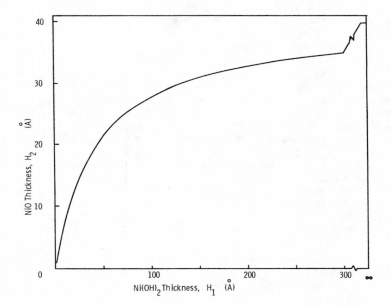

Figure 5. Thickness of NiO platelets vs. thickness of Ni(OH)₂ as calculated from Equation 6

9. Smith, J.V., Ed., "X-ray Powder Data File," ASTM Phila-
    delphia, 1967.

10. Helms, W.R., Mullen, J.G., Phys. Rev. B, (1971), $\underline{4}$, 750.

17

The Measurement of Low Interfacial Tension *via* the Spinning Drop Technique

J. L. CAYIAS, R. S. SCHECHTER, and W. H. WADE

Departments of Chemistry and Chemical Engineering, The University of Texas at Austin, Austin, Tex. 78712

Introduction

A method for obtaining interfacial and surface tensions by measuring the shape of a drop of liquid or bubble of gas in a more dense liquid contained in a rotating horizontal tube was first suggested by Plateau's (1) experiments. Beer (2) in 1855 considered the problem of the shape of the drop in some detail, as did Lord Rayleigh (3) in 1914.

It was not until 1942, however, that Vonnegut (4) suggested the method be used for measuring surface and interfacial tensions. Vonnegut's method rested on the approximation that the drop could be treated as a cylinder with hemispherical ends and thus avoided having to solve exactly for the drop shape. The diameter of the cylinder was the sole dimension of interest used to calculate the tension. Silberberg (5) improved Vonnegut's method by calculating correction factors for low speeds or small volumes. Rosenthal (6) in 1962 considered the stability of such a rotating drop determining a minimum length stability requirement. He also used Rayleigh's approach and solved for the shape exactly by using elliptic integrals. However, the shape solution was not solved explicitly for the tension. Princen, Zia, and Mason (7) in 1967 also solved the shape problem exactly by using elliptic integrals but more importantly solved explicitly for the tension. Their treatment focused on developing a method for measuring the length of the drop rather than the diameter to determine the surface or interfacial tension, and necessitated measuring the drop volume as well. They were able to show that for an infinitely long drop the exact solution reduced to Vonnegut's solution. Both Princen et al. and Rosenthal found the cylinder with hemispherical ends approximation to the shape to be adequate when the length to diameter ratio of the drop was greater than 3.5. This was also in general agreement with the correction terms Silberberg has calculated. However, all authors point out that this approximation is dependent upon the accuracy desired, the accuracy increasing as the ratio increases.

Patterson, Hu, and Grindstaff (8) in 1971 used both the
Princen et al. and the Vonnegut methods for measuring tensions of
polymers, and pointed out that the advantage of measuring the
diameter rather than the length of the drop is that it is not
necessary to establish the volume of the drop in the infinite
length case. The cited disadvantages are the difficulties in
measuring precisely small diameters and optical correction effects. In this paper it will be shown that in measuring low
interfacial tensions the disadvantages of measuring the diameter
are small in comparison to establishing the volume precisely.
Ryden and Albertsson (9) in 1971 also used the Princen et al.
method for measuring the interfacial tensions of polymer two-
phase systems. They obtained tensions down to 5 x 10^{-4} dyne/cm
for three component systems close to the miscibility limit.
Measurements were obtained at speeds between 200 and 450 RPM.
The high viscosities and near equal density phases studied, per-
mitted these low speeds and thence large, accurately determinable
drop volumes. Such phase properties are not usually encountered.

Theory and Calculations

The authors' method using the spinning drop technique, dif-
fers from the methods of Princen et al. and Vonnegut. Its theory
is based upon measuring both the length and width of the drop
rather than just one or the other and has the advantage that the
measurement of the volume is not necessary but can readily be
calculated.
The formulation of the problem is primarily that of Beer
(2), and the solution primarily that of Princen et al. (7). The
assumptions involved are: the angular velocity of rotation, ω,
of the drop is sufficiently large that the buoyancy effect due
to gravity is negligible; the axis of the drop is aligned on the
horizontal axis of rotation; the surface of the drop is described
by a surface of revolution; and the surface or interfacial tension, γ, is not a function of curvature. Rectangular cartesian
coordinates x, y, z are chosen (Figure 1) with the origin at the
left-hand end of the drop. The semiaxes of the drop are x_0 and
y_0. The densities of the drop and the outer phase are d_1 and d_2
($d_2 > d_1$) respectively. The radius of curvature of the drop surface at the origin is ρ_0.
The pressure difference ΔP across the interface of the drop
is given by

$$\Delta P = \gamma \left(\frac{1}{\rho_{xy}} + \frac{1}{\rho_{yz}} \right) = \frac{2\gamma}{\rho_0} - \frac{\Delta d \omega^2 y^2}{2} \tag{1}$$

where $\Delta d = d_2 - d_1$, and ρ_{xy} and ρ_{yz} are the radii of curvature in
the xy and yz planes, respectively.
The radii of curvature are given as:

$$\frac{1}{\rho_{xy}} = \frac{- d^2y/dx^2}{[1 + (dy/dx)^2]^{3/2}} \tag{2}$$

$$\frac{1}{\rho_{yz}} = \frac{1}{y[1 + (dy/dz)^2]^{1/2}} \tag{3}$$

Combining Equations (1), (2), and (3) with some mathematical manipulation yields

$$\frac{1}{y}\frac{d}{dy}\left(\frac{y}{[1 + (dy/dx)^2]^{1/2}}\right) = \frac{2}{\rho_o} - \frac{\Delta d\omega^2 y^2}{2\gamma} \tag{4}$$

which can be written in the dimensionless form

$$\frac{d}{dY}\left(\frac{Y}{[1 + (dY/dX)^2]^{\frac{1}{2}}}\right) = 2Y - \alpha Y^3 \tag{5}$$

where $Y = y/\rho_o$, $X = x/\rho_o$, and

$$\alpha = \frac{\Delta d\omega^2 \rho_o^3}{2\gamma} = 2C\rho_o^3 \tag{6}$$

$$C = \frac{\Delta d\omega^2}{4\gamma} \tag{7}$$

upon integrating Equation [5] one has

$$\frac{Y}{[1 + (dY/dX)^2]^{\frac{1}{2}}} = Y^2 - \frac{\alpha}{4}Y^4 \tag{8}$$

which can be reformulated to yield

$$\frac{dY}{dX} = \left[\frac{1}{Y^2(1 - \frac{\alpha}{4}Y^2)^2} - 1\right]^{\frac{1}{2}} \tag{9}$$

For $0 \le \alpha \le \frac{16}{27}$ Equation (9) can be integrated to obtain

$$X_o = \int_0^{Y_o} \frac{Y(1 - \frac{\alpha}{4}Y^2)dY}{[1 - Y^2(1 - \frac{\alpha}{4}Y^2)^2]^{\frac{1}{2}}} \tag{10}$$

To establish the bounds on α for the validity of Equation (10) one needs to consider Equation (8) evaluated at $Y = Y_o$. At this point $\frac{dY}{dX} = 0$ and one has

$$\alpha Y_0^3 - 4Y_0 + 4 = 0 \tag{11}$$

One root of this equation gives a physically acceptable value of Y_0 as a function of α.

Next, consider Equation (5) at $Y = Y_0$. For a very long drop (i.e. infinite length case) $\frac{1}{\rho_{xy}} = 0$ and $\frac{dY}{dX} = 0$ so Equation (5) reduces to:

$$\alpha Y_0^3 - 2Y_0 + 1 = 0 \tag{12}$$

Hence for a long cylindrical drop combining equations (11) and (12) one has

$$Y_0 = 3/2 \tag{13}$$

and the largest possible value for α of

$$\alpha = 16/27 \tag{14}$$

Combining Equations (6), (13), and (14) leads to Vonnegut's equation

$$\gamma = \frac{\Delta d \omega^2 Y_0^3}{4} \tag{15}$$

Convenient numerical integration of Equation (10) requires the following substitution:

$$q = 1 - \frac{\alpha Y^2}{4} \tag{16}$$

$$q_1 = 1 - \frac{\alpha Y_0^2}{4} \tag{17}$$

which gives

$$X_0 = \frac{1}{\sqrt{\alpha}} \int_{q_1}^{1} \frac{q \, dq}{[q^3 - q^2 + \frac{\alpha}{4}]^{\frac{1}{2}}} \tag{18}$$

If $q_1 > q_2 > q_3$ are the three roots of the cubic term in the denominator then

$$q_1 = 1 - \frac{\alpha}{4} Y_0^2 = \frac{2}{3} \cos \frac{\phi}{3} + \frac{1}{3} \tag{19}$$

$$q_2 = \frac{2}{3} \cos \left(\frac{\phi}{3} + \frac{2\pi}{3} \right) + \frac{1}{3} \tag{20}$$

$$q_3 = \frac{2}{3} \cos \left(\frac{\phi}{3} + \frac{\pi}{3}\right) + \frac{1}{3} \tag{21}$$

where
$$\cos \phi = 1 - \frac{27}{8}\alpha \tag{22}$$

and the solution of Equation (18) as given by Grohner and Hofreiter (10) for $q_1 \le q \le 1$ is

$$X_0 = \frac{2}{\sqrt{\alpha(q_1 - q_3)}} \left[q_1 F(k,\phi) - (q_1 - q_3)E(k,\phi) \right.$$

$$\left. + (q_1 - q_3)(\tan \phi) \sqrt{1 - k^2 \sin^2 \phi} \right] \tag{23}$$

where F and E are elliptic integrals of the first and second kind defined by:

$$F(k,\phi) = \int_0^{\phi} \frac{d\theta}{\sqrt{1 - k^2 \sin^2 \theta}} \tag{24}$$

and
$$E(k,\phi) = \int_0^{\phi} \sqrt{1 - k^2 \sin^2 \theta} \, d\theta \tag{25}$$

where
$$k^2 = \frac{q_2 - q_3}{q_1 - q_3} \tag{26}$$

and
$$\phi = \text{Arcsin} \left[\frac{1 - q_1}{1 - q_2}\right]^{\frac{1}{2}} \qquad 0 \le \phi \le \frac{\pi}{2} \tag{27}$$

The volume of the drop is obtained by taking Equation (8) in the form

$$\frac{1}{[1 + (dY/dX)^2]^{\frac{1}{2}}} = Y[1 - \frac{\alpha}{4}Y^2] \tag{28}$$

If one differentiates this equation with respect to Y and integrates with respect to X one has

$$\int_0^{X_0} \frac{d}{dY} \left\{ \frac{1}{[1+(dY/dX)^2]^{\frac{1}{2}}} \right\} dX = X_0 - \frac{3\alpha}{4} \int_0^{X_0} Y^2 dX \tag{29}$$

The left hand side of Equation (29) can be integrated if the substitution $\frac{dY}{dX} = \tan T$ is made. This yields

$$\int_0^{\frac{\pi}{2}} \cos T \, dT = 1 \tag{30}$$

and

$$\frac{V}{\rho_0^3} = 2\pi \int_0^{X_0} Y^2 dX = \frac{4\pi}{3} \left(\frac{r}{\rho_0}\right)^3 \tag{31}$$

where r is the radius of a sphere of the same volume as the drop. Then combining Equations (29),(30), and (31) yields

$$\frac{V}{\rho_0^3} = \frac{4\pi}{3} \left(\frac{r}{\rho_0}\right)^3 = \frac{8\pi}{3\alpha} (X_0 - 1) \tag{32}$$

This can be reduced to

$$\frac{r}{\rho_0} = \left[\frac{2(X_0 - 1)}{\alpha}\right]^{\frac{1}{3}} \tag{33}$$

Combining Equations (6) and (33) yields

$$Cr^3 = \frac{\alpha}{2} \left(\frac{r}{\rho_0}\right)^3 = X_0 - 1 \tag{34}$$

which will be an important equation in computing γ. With the aid of a high speed computor (CDC 6600), one can construct a table of values for r/ρ_0, Cr^3, x_0/r, y_0/r, and x_0/y_0 with α as the independent variable by the following scheme:

1. Assign value to α.

2. Calculate Y_0 by Equation (11) or Equation (19) for $0 \le Y_0 \le \frac{3}{2}$

3. Calculate X_0 by Equation (23).

4. Calculate $\frac{r}{\rho_0}$ by Equation (33).

5. Calculate Cr^3 by Equation (34)

6. Calculate x_0/r by $X_0 \div (r/\rho_0)$.

7. Calculate y_0/r by $Y_0 \div (r/\rho_0)$

8. Calculate x_0/y_0 by $X_0 \div Y_0$

   This produces Princen's table except the authors have expanded and extended it to $x_0/y_0 = 4.0$ ($\alpha = .592589328$), and because of the ease of extracting full accuracy from a computer,

the values are recorded to ten significant figures, this
accuracy however is not necessary.*
     The authors' method of using the table to obtain a value
of γ is to take the ratio of the length to width ($2x_0/2y_0$ =
x_0/y_0) of the drop and determine the parameter C by using the
table; then calculate γ from Equation (7). A flow diagram of
the procedure is as follows:

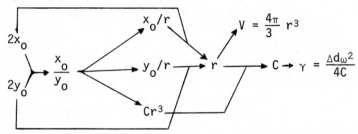

     When the ratio of the length to width is greater than 4.0,
the authors assumed that the shape can be adequately described
by assuming a cylindrical tube with hemispherical ends. This is
Vonneguts method and equation (15) can be used to determine γ
with less than 0.05% error. The error in calculating the volume
is less than 0.1%, using this geometry.

Experimental

     The apparatus used (Figure 2) is a refined version of that
employed by Princen et al. A hysteresis synchronous motor was
used. Its speed was controlled by varying the frequency from a
frequency generator. The rotational stability was 1 part in 10^5
as determined by a period averaging counter stable to 1 part in
10^8. The range of speeds used was from 1,200 RPM to 24,000 RPM.
A Gaertner traveling microscope with a filar eyepiece was used to
measure the length and width of the drop and to calibrate the
glass tubes for their magnification of the drop diameter. This
effect was determined to be a constant for the experiment of
y measured/y true=1.332 for all aqueous phases studied. The
tube housing and assembley (Figure 3) was designed to accept a
precision ground .245"O.D. pyrex glass tube (C) rounded on one
end and sealed against a rubber septum (G) on the other. A glass
cell enclosed the apparatus to permit temperature studies.
Thermostating of the system to $\pm$ 0.5C° was obtained.
     The procedure for loading the cell is to fill the glass tube
and metal cap (H) completely with the more dense phase. Holding
the tube upside down (capillary pressure retains the more dense
phase in the tube), the less dense phase is then injected with a
microliter syringe, the tube is then placed in the cap and the

* A copy of the program is available from the authors on request

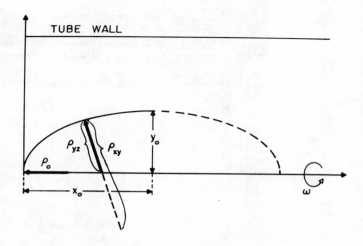

Figure 1.   Plane section of rotating drop with coordinate system imposed to describe the shape

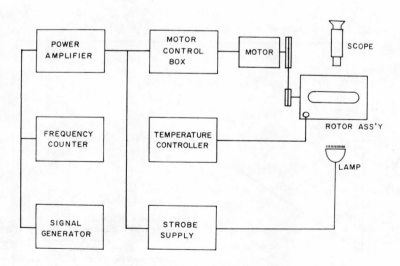

SPINNING-DROP   APPARATUS

Figure 2.   Schematic of spinning-drop apparatus

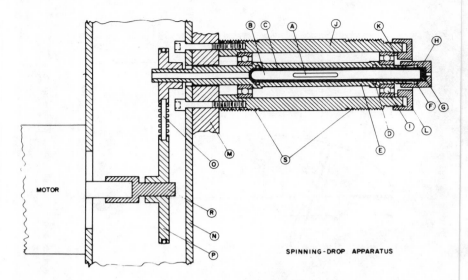

SPINNING-DROP APPARATUS

Figure 3. A = less dense phase (drop), B = more dense phase, C = glass tube, D = shaft O-rings, E = shaft, F = cap O-ring, G = silicone rubber septum, H = cap, I = ball bearings, J = outer support housing, K = O-ring for loading bearings, L = load adjusting cap, M = Bakelite plate, N = aluminum box tubing, O = non-slip positive drive belt, P = pully, R = motor shaft adaptor, and S = threads for heater wire windings

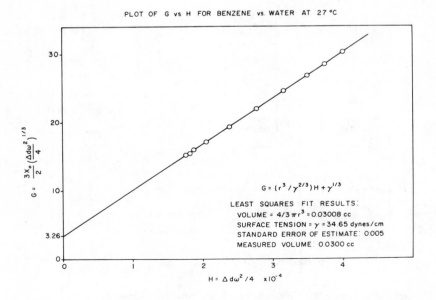

Figure 4.   Plot for obtaining α using the alternate Princen et al. method

whole assembly placed in the shaft and secured by screwing the cap
onto the shaft. A small hypodermic needle is inserted into the
cell through the septum to release any pressure buildup caused by
screwing the cap on. This method, though difficult at first, is
easily mastered after a few attempts.

In a test of the equipment using the benzene/water system
with a literature value (11) for the interfacial tension of 35.0
dynes/cm at 20°C the authors obtained at 27°C 34.50 dynes/cm using
the principal Princen et al. method of length and volume measurement and 34.65 dynes/cm using the alternate Princen et al. method
of frequency differences (Figure 4). Figure 5 is a photograph of
the drop. The slightly lower values could possibly be due to impurities in the system or the higher temperature or both. As a
final test of the system, the butanol/water system was measured
and a tension of 1.80+.03 dynes/cm was obtained at 27°C which compares very well with the literature value (11) of 1.80 dyne/cm
at 20°C.

To check the applicability of the system for measuring low
interfacial tensions a commercial surfactant, Petronate TRS 10-80
obtained from Witco Chemical Company, was used. A stock solution
was made up with the following components in water: 0.2 wt. %
Petronate and 1.0 wt. % NaCl. This solution was then compared
against several hydrocarbons at 27°C. The results of this experiment are in Figure 6. A picture of one of these drops is
included in Figure 7.

The octane/surfactant system verified the capability of the
apparatus to measure low tensions. The reproducibility of the
measurements was 2% but a confidence limit of 10% is probably more
realistic for tensions in this range. Using the 10% confidence
limit, tensions of 10^{-6} dyne/cm could be measured with the
Gaertner filar eyepiece (accurate to 5 X 10^{-5} cm) with some
difficulty. This is not a lower limit of the method, only of the
optical equipment used. The limit could possibly be extended by
choosing a different measuring system.

The main difficulty encountered was to deliver small volumes
of 10^{-3} cc or less. The discharge as a single drop of such small
volumes from a syringe was found to be very difficult. The drop
has a tendency to stay attached to the needle and quick removal of
the needle from the more dense phase caused the drop to detach
itself from the needle partially. This partial detachment made
accurate volume measurements almost impossible. A further
complication of the problem was that as the needle was retracted
from the liquid, some small drops of the less dense phase would
be left behind and after starting the rotation these small drops
could coalesce with the large drop changing its volume. Another
difficulty with the measured volume method is if the drop breaks
up after injection then the experiment would have to be term-
inated. The last difficulty with measuring the volume of the drop
was volume change due to solubilization. For low interfacial

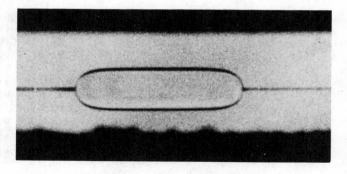

Figure 5.   Benzene water system at 20,000 RPM

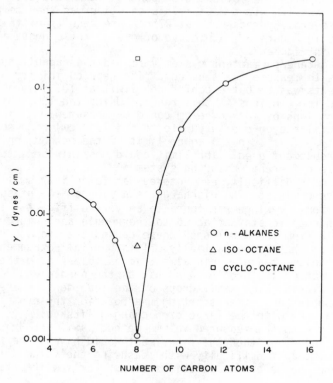

Figure 6.   Results of hydrocarbon studies against the surfactant

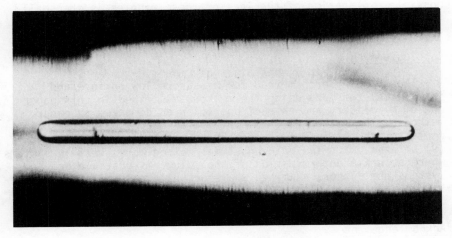

Figure 7.   Octane surfactant system at 6,000 RPM

tensions, where the drop was composed of several components, solubilization could result in indeterminate drop composition. For example the volume of the drop in the octane/surfactant system when the two phases were not pre-equilibrated, initially increased and then slowly decreased from the initial volume making the measured volume method useless and the frequency difference method inaccurate since the volume was not constant. For these reasons it was necessary to abandon the Princen et al. methods and use the authors' method described previously of measuring both the length and diameter of the drop and from this calculating the tension.

Both Vonnegut and Princen et al. observed a rippled surface for an air buble in water. Princen et al. suggested it was probably due to the vibration from the motor drive, however the authors in examining this aspect found no distortions of the surface but had vibrations present from the motor drive. A possible explanation of the lack of these ripples is the small size of the authors' system in comparison to that of Princen et al.

Conclusions

The spinning drop method for measuring low surface and interfacial tensions using the methods described in this paper is one of simplicity and ease of operation as compared to other methods. The method does not involve a third (solid) phase in contact with the interface of the drop as Ryden and Albertsson (9) pointed out. The Princen et al. method works well in systems where accurate determination of the volume is possible but for systems the authors examined where accurate determination of the volume is a difficult task, the method of measuring both the length and width of the drop is more accurate.

Acknowledgement

The authors wish to express their appreciation to the Robert A. Welch Foundation and the National Science Foundation for sponsoring this research and their special thanks and appreciation to H.M. Princen for valuable discussions and furnishing the authors with some of his original data. The authors also wish to express their gratitude to James Gardner for the drawings which appear in the text.

Literature Cited

1. Plateau, J.A.F., "Statique Experimentale et Theorique des Liquides, etc.", Gauthier-Villars, Paris, 1873.

2. Beer, A., Annalen der Physik und Chemie von Poggendorff, (1855) 96, 210.

3. Raleigh, Lord, Phil. Mag., (1914) <u>28</u>, 161.

4. Vonnegut, B., Rev. Sci. Instrum. (1942) <u>13</u>, 6.

5. Silberberg, A., Ph.D. Thesis, Basel University, Switzerland, 1952.

6. Rosenthal, D.K., J. Fluid. Mech., (1962) <u>12</u>, 358.

7. Princen, H.M., Zia, I.Y.Z., and Mason, S.G., J. Colloid Interface Sci., (1967) <u>23</u>, 99.

8. Patterson, H.T., Hu, K.H., and Grindstaff, T.H., J. Polymer Sci., (1971) <u>34</u>, 31.

9. Ryden, Jan and Albertsson, Per-Åke, J. Colloid Interface Sci. (1971) <u>37</u>, 219.

10. Grobner, W. and Hofreiter, N., "Integraltafel," 3rd ed., Vol. I, p. 78, Springer Verlag, Vienna, 1961.

11. Adamson, Arthur W., "Physical Chemistry of Surfaces," 2nd ed., pp. 44-45, Wiley, New York, 1967.

18

Adhesion of Ice Frozen from Dilute Electrolyte Solutions

H. H. G. JELLINEK

Department of Chemistry, Clarkson College of Technology, Potsdam, N. Y. 13676

Introduction

The adhesion of ice has long been studied but a satisfactory solution, from the practical standpoint of decreasing adhesion or increasing abhesion, has not yet been found. Complete abhesion can be achieved under laboratory conditions but in practice any surface becomes rapidly contaminated after a few abhesions have taken place and the adhesion of ice continually increases with further abhesions.

Experimental Data by Smith-Johannsen and Discussion

Smith-Johannsen ([1]) studied the effect of impurities in water on the adhesion of ice by freezing such solutions to various substrates. He ascertained that small initial concentrations (1×10^{-3} mole/liter(M) or less) of electrolytes (salts) decrease ice adhesion considerably. The adhesive strength was measured by the force per square centimeter needed to shear the ice off the substrate. The experiments were carried out at $-10°C$. The freezing point lowering of water by electrolytes of such concentrations is only about $0.005°C$. The ice was frozen rapidly enough so that the electrolyte remained homogeneously distributed throughout the frozen system, and did not have sufficient time to diffuse away from the substrate/solution interface. Air bubbles appeared during the freezing process. Table I gives relevant data (S = salinity of initial solution; p = salinity of grain boundary solution).

The effectiveness of decreasing adhesion, or increasing abhesion, varies considerably with the particular electrolyte in solution. Thus $1 \times 10^{-3}M$ $Th(NO_3)_4$ decreases adhesion by 97% on a wax-treated aluminum

248

Table I.   Relevant Data for Adhesion of 10^{-3} M Electrolyte Solutions Frozen to a Wax-treated Aluminum Surface at -10°C.

Electrolyte	$10^4 \frac{s}{p}$	$10^{-3}\frac{p}{s}$	Adhesive strength A (10^{-3} g/cm²)	$D_0^{1/2}(0°C)*$ (10^3 cm/sec$^{1/2}$)	$m=d(T_{abs})/d(\%\ Comp)*$
NaCl	3.44	2.91	2.45	2.79	0.67
KCl	3.24	3.08	2.20	3.15	0.90
Na-acetate	5.13	1.95	0.85	2.36	0.94
NaNO₃	2.50	4.00	0.65	2.78	0.29(?)
K-acetate	5.17	1.93	0.76	2.57	0.98
NH₄Cl	3.34	2.99	2.78	3.15	0.66
CaCl₂	8.54	1.17	0.39	2.63	1.27
Ca(NO₃)₂	7.13	1.40	1.38	2.62	0.65
MgCl₂	8.66	1.15	1.30	–	–
Th(NO₃)₄	11.14	0.88	0.11	–	–
Ca(NO₃)₂	6.48	1.54	<0.10	–	–
Ni(NO₃)₂	8.32	1.20	<0.10	–	–
Be(NO₃)₂	7.86	1.27	<0.10	–	–
KCNS	4.62	2.17	<0.10	–	–

*The diffusion coefficients D_0 were calculated from the Nernst equation using data from the Handbook of Physics and Chemistry, 48th Edition, Chemical Rubber Publishing Co. The liquidus slopes and p values were derived from data in A. Seidall, Solubilities of Inorganic and Metal Organic Compounds, 3rd Edition, D. van Nostrand Co., Inc., 1940. m is the slope of the liquidus curve in the respective phase diagram i.e. $m = d(T_{abs})/d$ %(Composition).

surface, whereas the same molar concentration of NaCl causes a decrease   of only 42%. The term adhesion is not used here in its strict sense, i.e., a break in the substrate/ice interface; any breaks very near the interface are included in this term.

With a few exceptions, the total salt concentration eventually has to accumulate in grain boundaries when such a solution freezes; the concentration of the electrolyte is practically zero in the lattice of the ice grains. The grain boundary consists of a saturated solution in accordance with the phase diagram of the respective electrolyte, provided the temperature is above the eutectic temperature. The width of the grain boundary can be calculated if the phase diagram is known; this has been done, for instance, by Chatterjee and Jellinek (2) for NaCl solutions.

Smith-Johannsen (1) pointed out that effective electrolytes should have endothermic heats of solution,[*] low eutectic temperatures and appreciable solubility at 0°C. If the solution is frozen below the eutectic temperature, then the adhesion increases considerably.

It is of interest to consider the possible conditions at or near the substrate/ice interface. This may give a clue as to the mechanism of the reduction of adhesion by electrolytes initially present in dilute solutions. If it is assumed that a thin liquid solution layer is formed between the ice and substrate, then the ice could be relatively easily sheared off the substrate. The shear force per square centimeter or the adhesive strength would depend on the thickness and viscosity of the layer and the rate of shear. The formation of such a layer is assumed to be subject to the same laws as the formation of grain boundaries but the layer would not necessarily be of similar thickness. The latter may also be a function of the nature of the substrate.

The width or thickness δ of a grain boundary for prismatic ice grains of quadratic cross section is given by (2)

$$\delta = \frac{s\rho_{i,T}\bar{b}}{2p\rho_{gb,T}} \qquad (\delta << \bar{b}) \qquad (1)$$

where $\bar{b}$ is the average width of the grains, i.e., the

---

*It is doubtful that this is necessary; the adhesion value for $FeCl_3$, which has an exothermic heat of solution, fits the value for HCl well. $FeCl_3$ is practically completely hydrolyzed.

length of the edge of the quadratic cross section, s
and p are the salinities (in g/1000g of solution) of
the initial solution and the saturated grain boundary
solution, respectively, at T K, and $\rho_{i,T}$ and $\rho_{gb,T}$ are
the densities of ice and the grain boundary at T K,
respectively. The width for an interfacial solution
layer can be multiplied by a factor to take care of the
characteristics of the substrate. The right side of
Equation (1) has to be multipled by $^2/_3$ if the grains
are spherical.

The adhesive strength A for this case is given by

$$A = \frac{\eta V}{\delta} \qquad (2)$$

where η is the viscosity of the layer and V is the
shear velocity. The layer is assumed to behave in a
Newtonian fashion; there is no evidence that such solution layers behave otherwise.

Introducing Equation (1) into Equation (2) yields

$$A = \frac{2\eta V\rho_{gb,T}}{\rho_{i,T}\bar{b}} \cdot \frac{p}{s} \,. \qquad (3)$$

The ratio $\rho_{gb,T}/\rho_{i,T}$ and the viscosity vary relatively little with the nature of the electrolytes and
their concentrations; $\bar{b}$ depends mainly on the rate of
freezing, and V and T are kept constant. Hence, the
adhesive strength will be predominantly influenced by
the ratio p/s. Equation (3) can then be simplified to

$$A = K \frac{p}{s} \qquad (4)$$

where K is a constant. Figure 1 shows A plotted versus
p/s using data from Table I. Equation (4) is as well
obeyed as can be expected under the circumstances. All
solutions were initially 1×10^{-3}M. Hence, in order to
obtain s in g/1000 g of solution, the molarity has to
be multiplied by the molecular weight of the respective
electrolyte. The straight lines and all subsequent
ones have been drawn by the method of least squares.

However, this situation at the interface considered above is not the only one which can be visualized.
The ice prisms could be directly frozen with their
quadratic cross sections to the substrate. These
prisms would then be separated from each other by grain
boundaries of width according to Equation (1). The

adhesive strength in this case can be assumed to be given by the force per square centimeter needed to rupture by shear the ice prisms contained in 1 cm^2 of interface, hence

$$A = A_0 (aY)^2 \tag{5}$$

where A_0 is the force necessary to rupture ice with an actual cross section of 1 cm^2, a is the number of prismatic grains along 1 cm in the frozen system, and Y^2 is the cross-sectional area of one prism. It has to be noted that all dimensions are average quantities in this context. The term $(aY)^2$ is equal to the fraction of 1 cm^2 occupied by the prisms, i.e. it is a pure number.

Y can easily be related to δ, the width of the grain boundary:

$$aY + (a-1)\delta = a(Y + \delta) = 1. \tag{6}$$

Hence, introducing δ from Equation (6) into Equation (5) gives

$$A = A_0 (1-a\delta)^2. \tag{7}$$

Further, introducing Equation (1) into Equation (7), assuming again that all parameters except p and s vary little, yields

$$A = A_0 (1 - K' \frac{s}{p})^2. \tag{8}$$

Hence, $A^{1/2}$ plotted versus s/p should result in a straight line. This is as satisfactorily the case as in the previous instance (see Figure 2).

However, a difficulty arises here. The grain boundary and the interfacial layer widths are quite small, hence even doubling the grain boundary width should decrease the adhesive strength by only a small amount. However, in actual fact the decreases in adhesive strength are quite large. Hence, Equation (5) or (7), respectively, cannot be correct. Equation (7) should rather be of the form

$$A = A_0 (1 - a^n \delta)^2 \tag{9}$$

where the exponent n is larger than one. This results in the following functional relation between A and Y:

$$A = A_0 [1 + a^{n-1}(aY - 1)]^2. \tag{10}$$

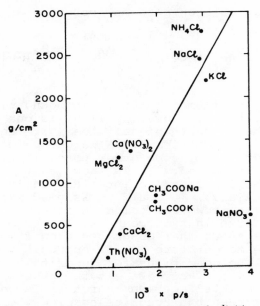

Figure 1.   Adhesive strength vs. ratio of salinities
in grain boundaries to those in initial solutions
($-10°C$, 10^{-3}M solutions)

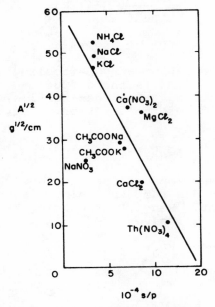

Figure 2. Square root of adhesive
strength vs. ratio of initial to grain
boundary salinity

Equation (9) can also be expressed as

$$A = A_0 \left[1 - \frac{\nu k}{h^{1/2}} \right] = A_0 \left[1 - \frac{a^n \nu^{1/2}}{h^{1/2}} \right]^2 \qquad (10a)$$

where ν is the volume of the grain boundaries between prisms of height h. Equation (10a) is one of the forms for the strength of salt ice (3). Equation (8), however, is still valid with a larger value of K'.

There are still some other possibilities which have to be considered. It is known that dendrites are formed at the solution/ice interface during freezing of salt solutions. It could be assumed that small dendrites also grow from solid surfaces under suitable conditions. (See in this connection Bascom et al. (4).) Camp (5) has shown that dendrites grow from solid surfaces if the temperature is low enough. French (6) has deduced the conditions for dendrite growth at the solution/ice interface. If his theory is applied to the solid/ice interface (whether this is permissible is doubtful) then the dendrite spacing L, i.e. the distance between the centers of two adjacent dendrites, can be calculated. The relevant expression is

$$L = \frac{4 \Delta T \sqrt{D_0 \Theta_f}}{C_0 \sqrt{\pi}} \qquad (11)$$

where ΔT is the extent of supercooling (for details see reference 6), D_0 is the diffusion coefficient of the electrolyte, Θ_f the freezing time, and C_0 the initial molality of the solution. Further, French (6) derived that

$$\Delta T = -mC_0 \beta \sqrt{\pi} \qquad (12)$$

where m is the slope of the liquidus curve in the respective phase diagram (this slope is negative) and β is a constant. Hence, introducing Equation (12) into Equation (11) gives

$$L = -4m\beta \sqrt{D_0 \Theta_f} \qquad (13)$$

D_0 varies very little compared with m. If dendrites are assumed to have the shape of quadratic prisms, then the adhesive strength is again given by Equation (10). L is related to Y as follows:

$$(a-1)L + Y = 1. \tag{14}$$

Hence Equation (5) gives

$$A = A_0 a^2 [1 - (a-1)L]^2. \tag{15}$$

Introducing Equation (13), assuming everything to be constant except L, gives

$$A = A_0 a^2 (1 + K'' m)^2. \tag{16}$$

The same difficulty arises here as in the previous case. L is very small and the decrease of A with L is much greater than is possible for small values of L. Hence (a-1) can be raised to a power n; this does not affect Equation (16) except for the magnitude of K''. Equation (15) then becomes

$$A = A_0 a^2 [1 - (a-1)^{n-1}(1-Y)]^2. \tag{17}$$

$A^{1/2}$ plotted versus m again gives a satisfactory straight line (see Figure 3).

Rohatgi and Adams ([7]) made calculations similar to those of French ([6]). Their results can be expressed by an equation as follows:

$$L^2 = 8D \frac{p}{s} \frac{d\Theta}{dfs} \tag{18}$$

where $d\Theta/dfs$ is the reciprocal rate of freezing (fs = fraction of solid). Equation (18) leads to a contradiction. It indicates that L increases with p/s; in other words δ decreases with increasing values of p/s, which contradicts Equation (1). Equation (18) can only become consistent with Equation (1) if $d\Theta/dfs$ changes so that the increase in p/s is more than compensated. This inconsistency has already been pointed out by Lofgren and Weeks ([8]).

Thus the above discussion shows clearly that the variations in grain boundary or in interfacial solution layer thickness are mainly responsible for the variations in adhesive strength. However, a decision cannot be made on the basis of the above considerations whether ice (dendrites or prisms) is ruptured

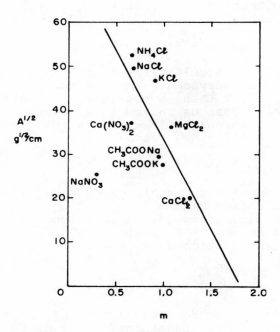

Figure 3.   Square root of adhesive strength vs. *slope of liquidus curve of phase diagram*

or whether an interfacial liquid solution layer is sheared.

Smith-Johannsen ([1]) also performed some experiments on the variation of adhesive strength with solute concentration. The results of these experiments give clues to the mechanism of the adhesion process. Table II gives adhesion values as a function of solute concentration for $Th(NO_3)_4$ and $CaCl_2$ solutions.

A plotted versus concentration gives curves which show very rapid initial decreases in adhesive strength with increasing electrolyte concentrations, which eventually slow down, resulting in straight lines with moderate slopes at concentrations near $5 \times 10^{-4}M$. Plots of $A^{1/2}$ versus concentration still show appreciable curvatures, as do plots of A versus reciprocal concentration.

The plot of A versus reciprocal initial concentration should result in a straight line according to Equation (3). Neither are Equations (8) and (15) obeyed; the latter should be taken in conjunction with Equation (11). The deviations from straight lines according to Equation (8), which has been derived on the assumption that ice prisms are ruptured, can, in principle, be accounted for by the known fact that the number of defects, dislocations or impurities in the ice lattice decreases in an ice prism with its size (in this case with the grain size). This could result in deviations from a straight line of a plot of A versus concentration as indicated by the experimental data. Also the experimental results evaluated by Equation (14) in conjunction with Equation (11) show deviations in the right direction.

Adhesive strength due to the interfacial liquid solution layer should be a linear function of the reciprocal initial concentration; however, as already indicated above, such plots have a very pronounced curvature, especially in their initial parts. The thickness of such layers over the whole range of concentrations is actually very small. Thus, it is conceivable that the roughness of the substrate surface will strongly influence the sliding of the ice across the substrate surface due to shear forces. This was found to be the case with thin water layers ([9]). Sliding of glaciers promoted by an interfacial water film is also of interest in this connection ([10]). Actually, the reciprocal sliding velocity as a function of water layer thickness between the ice and the bed of a glacier gives curves similar to plots of adhesive strength versus concentration observed by Smith-Johannsen ([11]) It is difficult to see why the whole base of the ice

Table II. Effect of Salt Concentration on Adhesive Strength (-10°C) (1)

Conc (mol/l)	Adhesive strength A (g/cm^2)	Change in adhesive strength (%)
Th(NO$_3$)$_4$, wax-treated aluminum surface		
0	4250	0
1 X 10^{-5}	3225	25
1 X 10^{-4}	1990	53
4 X 10^{-4}	1075	74
1 X 10^{-3}	110	97
CaCl$_2$, clean aluminum surface		
0	8000	0
1 X 10^{-4}	3300	59
7 X 10^{-4}	1800	78
1 X 10^{-3}	1600	81
2.5 X 10^{-3}	1100	87
5 X 10^{-3}	480	94
1 X 10^{-2}	130	98

prism should be frozen directly to the substrate.
There is no reason to suppose that grain boundaries
should not also exist between the prism and the substrate. Thus it seems more likely that an interfacial
liquid solution layer is sheared than that ice prisms,
directly frozen to the substrate, are ruptured.

The overall conclusion reached is that the variation of adhesive strength due to a number of electrolytes in dilute solutions is in large part due to the
shearing of an interfacial liquid solution layer which
is formed during freezing. The thickness of such a
layer may be influenced by the nature of the substrate.
This thickness is mainly proportional to the ratio of
the salinity of the initial solution to the salinity
of the saturated solution in the grain boundary. In
addition, equimolar solutions do not have identical initial salinities in terms of g/1000 g of solution; the
molarity has to be multiplied by the molecular weight
of the respective electrolyte. If the temperature of
the ice-substrate system lies below the eutectic point
of the electrolyte, then the adhesive strength is
appreciably increased compared with that for a temperature above the eutectic one.

The thickness of the interfacial layers is so
small for the range of 10^{-5} to about 10^{-3} mol/liter of
initial electrolyte concentrations that the roughness
of the substrate surface has an appreciable frictional
effect on the sliding velocity of the ice across the
substrate surface.

Although it cannot be concluded with certainty
that the above mechanism is the one which is actually
realized, it appears to be the most reasonable of the
possibilities that present themselves. Confirmation
of this mechanism could be obtained experimentally by
freezing dilute electrolyte solutions to optically
flat plates and observing the sliding velocity as a
function of shear force.

This work was supported by U.S. Army Corps of Engin-
eers, CRREL, Hanover, New Hampshire.

Summary

       Experiments by Smith-Johannsen on the adhesion of
ice frozen from a number of 1 X 10^{-3}M electrolyte solutions to a wax-treated aluminum surface at -10°C are
discussed. It is concluded that the adhesive strength
measured by the force per square centimeter needed to
shear the ice off the substrate surface is mainly due
to a liquid interfacial solution layer between the ice
and the substrate surface. The thickness of such a
layer is largely determined by the same considerations
as the thickness of grain boundary layers in ice obtained from dilute electrolyte solutions.

Literature Cited

1.   Smith-Johannsen, R., General Electric Company Re-
      port 5539, Pts. I and II (ATSC Contract W-33-038-
      AC-9151), pp. 149-191, 1946.
2.   Chatterjee, A.K. and Jellinek, H.H.G., Journal
      Glaciol.,(1971), 10, 293.
3.   Weeks, W.F. and Assur, A., in "Ice and Snow", W.D.
      Kingery, Ed., pp. 258-276, MIT Press, Cambridge,
      Mass., 1963.
4.   Bascom, W.D., Cottington, R.L., and Singleterry,
      C.R., J. Adhesion, (1969), 1, 246.
5.   Camp, P.R., Ann. N.Y. Acad. Sci., (1965), 125,
      317.
6.   French, D.N., Ph.D. Dissertation, Department of
      Metallurgy, Massachusetts Institute of Technology,
      Cambridge, Mass., 1962.
7.   Rohatgi, P.K. and Adams, C.M., Journal Glaciol,,
      (1967), 6, 663.
8.   Lofgren, G. and Weeks, W.F., Journal Glaciol.,
      (1969), 8, 153.
9.   Jellinek, H.H.G., U.S. Army Snow, Ice and Perma-
      frost Research Establishment (USA SIPRE) Special
      Report 37, 1960.
10.   Weertman, J., U.S. Army Cold Regions Research and
      Engineering Laboratory (USA CRREL) Research Re-
      port 162, 1964.
11.   Weertman, J., Can. J. Earth Sci., (1969), 6, 929.

Mechanism of Olfaction Explained Using Interfacial Tension Measurements

D. V. RATHNAMMA

The Ohio State University, 1680 University Drive, Mansfield, Ohio 44906

Introduction

Adsorption of the odorant on the receptor sites of the nerve fiber of the mucous membrane of the nose triggers a biochemical or biophysical mechanism which facilitates exchange of sodium and potassium ions which in turn fires a nerve impulse. If this is true there should be a relation between the threshold value of concentration of the vapor for giving detectable odor and adsorbability on mucous membrane. The physical model of olfactory apparatus used consisted of an interface formed by water and mineral oil. Interfacial tension was measured as a function of concentration of the odorant. The slopes of the curves of interfacial tension versus log concentration at an arbitrarily chosen small concentration of about 5×10^{-3} moles per liter seem to be proportional to the intensity of the odour. The concentration required to produce a given lowering in interfacial tension for n-alcohols decreases with increasing molecular cross section. The literature values for olfactory threshold concentrations for producing detectable odor also show decreases with increasing molecular cross section. Adsorption of some flavoring agents at the interface was also studied.

This paper contains the results of measurements of interfacial tension between mineral oil and water with various alcohols and odorants distributed between the two phases.

Materials

All the alcohols used were straight chain alcohols. The data presented in this paper were obtained using Baker Analyzed reagents. The alcohols used were ethanol, n-propanol, n-butanol, n-pentanol (amyl alcohol) and n-hexanol. Lauryl alcohol - Givaindan Delawanna, Inc., and n-hexadecanol - Brothers Chemicals, New Jersey, both of "manufacturing use only" quality were distilled before use.

Squibb mineral oil was also distilled and used without any further purification. The flavoring agents vanillin, cinnamic alcohol, eugenol, isoeugenol, safrol, cinnamic aldenhyde, phenyl propyl aldehyde, and phenyl propyl alcohol were samples supplied by the General Foods Corporation, Tarrytown, N.Y. The flavoring agents were used without any further purification.

Double distilled water distilled using quartz distillation unit was used for making up solutions of alcohols in water and also for the aqueous phase of the blank measurements.

Procedure

Solutions of each n-alcohol in distilled water were made up by dissolving weighed amounts of the alcohol in double distilled water and making up the solution to the required volume. Stock solution was made up and it was diluted with distilled water to lower concentrations.

100 ml of aqueous solution of n-alcohol formed the lower layer of the system in a beaker. The upper layer was 100 ml of Squibb mineral oil. Five minutes were allowed for the solute to distribute itself between the phases. Readings of interfacial tension measurements were constant showing that the time waited was sufficient to attain equilibrium.

Lauryl alcohol and n-hexadecanol are insoluble in water. These alcohols were dissolved in the oil phase and interfacial tension measurements were made by equilibriating the oil phase with distilled water for 5 minutes.

The temperature was controlled and maintained at 26 ± 0.1^oC by circulating thermostated water through the outer jacket of the beaker in which the oil/water system was placed. Any change in the mutual solubilities of the phases of the solubility of the alcohol in the phases due to this slight variation in temperature was small and could not have caused more than a few hundreths of a dyne/cm in the interfacial tension values (1).

Measurement of Surface and Interfacial Tension

The interfacial tension was measured by the wettable blade method which employs the Wilhelmy principle. The simplified Rosano tensiometer (2) used contains a sand blasted platinum blade of 5 cm perimeter. This Wilhelmy blade is dragged into the solution by the force of interfacial tension acting at the interface and the weight in milligrams applied to pull the blade to the surface is measured. This external force is applied from a torsion balance restoring the blade to the original position.

If R_w = meter reading for water
  γ_w = surface tension of water
  R_s = meter reading for the solution
  γ_s = surface tension of the solution

$$\gamma s = \frac{Rs}{Rw} \gamma w$$

$$\frac{\gamma w}{Rw} = K \qquad K \text{ is a constant.}$$

The constant K was evaluated by taking meter readings for water at two different temperatures. K was found to be 0.196. Therefore the meter reading for any solution if multiplied by 0.196 gives the value of the surface or interfacial tension.

The blade was flamed each time before use and was wetted with double distilled water since aqueous phase was involved. Explanation of the equation $\gamma_s = R_s \times .196$

$$\gamma_s = \frac{R_s \text{ (in mg)}}{w(\text{perimeter of blade in cm})} \times .980$$

γ_s = Surface tension (dynes/cm)

R_s = Meter readings in milligrams

$$\gamma_s = \frac{R_s \text{ (in mg)}}{5 \text{ cm}} \times (.980) = .196 (R_s)$$

Results and Discussion

The interfacial tension of the interface formed by aqueous solution of alcohol and Squibb mineral oil was measured for various concentrations of each alcohol in water. The data are presented in graphical form in Figure 1.

Table I. Adsorption Values From the Slopes of Curves of
Log c Versus Interfacial Tension

Alcohol	Slope at log c = -2.3	No. of molecules per square cm (adsorbed) Gibbs	Molecular cross section. A^2 from molecular volumes
Ethyl	.225	2.37×10^{12}	26.7
n. Propyl	.163	1.72×10^{12}	32.6
n. Butyl	.143	1.51×10^{12}	38.0
n. Amyl	.122	1.29×10^{12}	43.0
n. Hexyl	.103	1.09×10^{12}	47.8
Lauryl	.204	2.15×10^{12}	77.0
n. Hexadecanol	.275	2.90×10^{12}	97.0

Table II. Interfacial Tension Reduction 1 dyne/cm. Alcohol Concentration Required to Produce the Reduction.

Alcohl	Equalibrium conc of alcohol (aqueous) to produce Y reduction 1 dyne/cm M x 10^2 (from graph)	Experimental olfactory (3) threshold molecules/c.c.air
Ethyl	2.45	2.44×10^{15}
n. Propyl	2.69×10^{-1}	5.00×10^{13}
n. Butyl	3.99×10^{-2}	8.20×10^{12}
n. Pentyl	2.51×10^{-3}	6.80×10^{12}
n. Hexyl	4.27×10^{-4}	6.72×10^{12}
Lauryl	2.20×10^{-3}	
n. Hexadecanol	1.20×10^{-2}	

Table III. Flavoring Agents Conc. to Produce Interfacial Tension Reduction of 2 dyne/cm

Flavoring agent	Equilibrium Concentration (aqueous) M x 10^6
Vanillin	3.0×10
Cinnamic alcohol	1.85
Eugenol	9.5×10^{-1}
Isoeugenol	4.5
Safrol	3.15
Cinnamic aldehyde	7.0
Phenyl propyl aldehyde	2.0
Phenyl propyl alcohol	1.3×10

Olfactory threshold values for flavoring agents not available.

Calculation of Adsorption Using Gibb's Equation

$$d\gamma = \Gamma\ RTd\ \ln a$$

γ = Interfacial Tension

Γ = Surface excess per square centimeter

a = Activity

R = 8.31 x 10^7 ergs/deg/mole

T = 26^oC = 299^oK

If the slope = 1

$$\Gamma = \frac{1/cm^2}{8.31 \times 10^7/mole \times 299 \times 2.3} = \frac{1 \times 6 \times 10^{23}}{8.31 \times 10^7 \times 299 \times 2.3}$$

molecules/cm^2 = 1.054 x 10^{13} molecules/cm^2

It is believed (3,4,5,6,7,8)that the process of smelling involves adsorption of the odor producing material on to the surface of the mucous membrane of the nerve cell, producing a biochemical or biophysical mechanism to facilitate exchange of ions of sodium and potassium (3,6,8) to fire a nerve impulse. If this is the case then there should be a relation between the threshold value of concentration of the vapor for giving a detectable odor and the adsorbability on mucous membrane. Human nose is not a reliable instrument for measuring the threshold value, therefore different physical models have been employed by different workers. (9,10,11,12,13)

The surface excess according to Gibbs equation was calculated for each alcohol. The surface excess is highest for ethyl alcohol and decreases up to carbon atom 6. Then again the surface excess increases. Adsorption or surface excess parallels published (3)values for olfactory threshold and the adsorption decreases with increasing molecular cross section up to carbon atom 6 for n-alcohols.

The concentration of each alcohol solution in water required to reduce the interfacial tension of oil-water interface by 1 dyne/cm was determined. This concentration also decreased from alcohol with two carbon atoms to six carbon atoms. Thus these concentrations can be used to calculate olfactory threshold values.

If the oil-water interface is accepted as a simulated olfactory epithelium, the odorant molecules adsorbed at the interface at low concentration can be a measure of the olfactory threshold. In order to produce the sensation of odor a certain minimum reduction in the interfacial tension has to occur. For purposes of comparing different odorants 1 dyne/cm reduction is used in this investigation. This is equivalent to a decrease in the surface free energy of one erg per square cm.

The concentration of certain flavoring agents required to reduce the interfacial tension by about 2 dynes/cm at the oil/water interface was determined. With the exception of vanillin which was soluble in water, the others were soluble in Squibb mineral oil. Figure 2. No conclusions could be drawn since olfactory threshold values of flavoring agents are not available.

It was found that the extent of adsorption of the n-alcohols at the oil/water interface is related linearly to the published (3) threshold concentrations for the same alcohols. Figure 3.

Concentration of n-alcohol in the aqueous phase required to produce a lowering in interfacial tension of 1 dyne/cm also bears a linear relationship to the olfactory threshold values. Figure 4. These correlations show that the oil/water system is a satisfactory model for olfactory studies.

Summary of Olfaction Mechanism

It is generally agreed that a odor can be detected if the odor carrying substance makes physical contact with the interior part of the nose, namely, the membrane of the nerve cell. The human nose is very sensitive in detecting odorants in low concentrations and at low vapor pressures and the human nose is very selective. Simulated systems lack this kind of sensitivity and selectivity. Interaction between the stimulant and the receptor namely the trigeminal nerve endings takes place as a result of which stimulus is formed. The molecules of the stimulant or the products formed are removed from the interaction zone and the stimulus is transmitted to the olfactory region of the brain and translated into the sensation of odor. The possible mechanism is the odorant molecules interact with the nerve cells in the olfactory epithelium and the molecules are adsorbed on the lipid-water interface of the cell membrane. The odorant molecules penetrate the lipid-cell membrane which results in an exchange between potassium ions in the cell and the sodium ions outside the cell. This process triggers an impulse in the olfactory nerve. The oil/water interface can be likened to the lipid/water interface.

Literature Cited

1. Bikerman, J.J. "Physical Surfaces" pp.118-121. Academic press, New York, 1970
2. Salzberg, H.W.,Morrow, J.I., and Cohen, S.R. "Laboratory Course in Physical Chemistry" p. 103-104. Academic Press, New York, 1966.
3. Davies, J.T., Taylor, F.H. 2nd International Symposium on Surface Activity (1957) 4 Butterworth, London. pp.329-340.

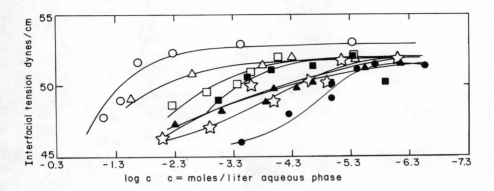

Figure 1.   *Interfacial tension vs. concentration for a variety of alcohols.* ○ = *ethanol,* △ = *n-propyl alcohol,* □ = *n-butanol,* ☆ = *amyl alcohol,* ● = *n-hexyl alcohol,* ▲ = *lauryl alcohol,* ■ = *n-hexadecanol.*

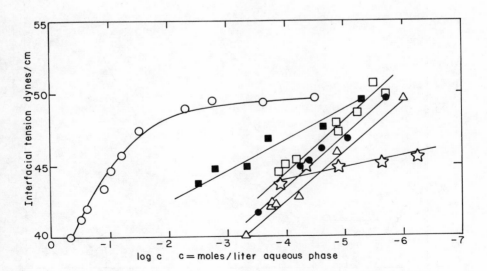

Figure 2.   *Interfacial tension vs. concentration for flavoring agents.* ○ = *vanillin,* ● = *cinnamic alcohol,* △ = *eugenol,* ■ = *isoeugenol,* □ = *safrol,* ☆ = *phenyl propyl alcohol.*

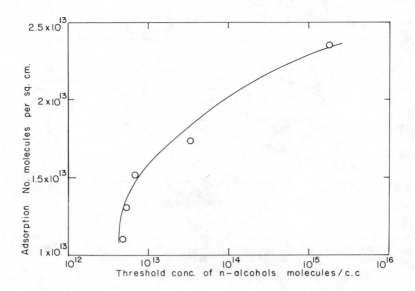

Figure 3. *adsorption* vs. *threshold concentration*

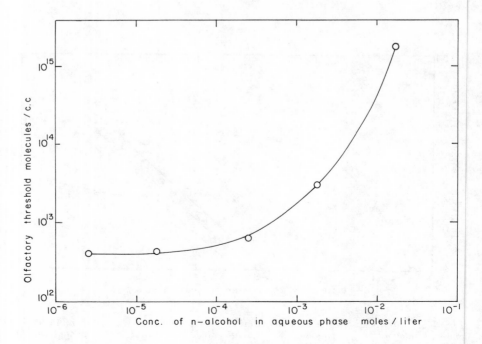

Figure 4. *Concentration of alcohol required to produce interfacial tension reduction of 1 dyne/cm vs. olfactory threshold*

4. Rosano, H.L., and Friedman, H.H., in "Flavor
    Chemistry," Adv. Chem. Ser. No. 56, pp. 53-63,
    American Chemical Society, Washington, D.C.,
    1966.
5. Beets, M.G.J., Mol., Pharmacol. (1964) 11, 1.
6. Davies, J.T., J. Theor. Biol. (1965) 8, 1.
7. Renshaw, E.H., Aust. J. Dairy Tech. (1964) 19(3),
    128.
8. Davies, J.T., "Recent Developments in the Penetra-
    tion and Puncturing Theory of Odour," Reprinted
    from the Ciba Foundation Symposium of Taste and
    Smell in Vertebrates, pp. 265-291, J an A
    Churchill, London, 1970.
9. Dravnieks, A., Ann. N.Y. Acad. Sci. (1964) 116,357.
10. Amoore, J.E., Proc. Sci. Sec. Toilet Goods Assoc.
    (October 1962) 37, 13.
11. Dravnieks, A., Ann. N.Y. Acad. Sci. (1964)116,429.
12. Hartman, J.D., Proc. Am. Soc. Hort. Sci. (1954)
    64, 335.
13. Wilkens, W.F., Hartman, J.D., J. Food Sci. (1964)
    29(3), 372.
14 Harkins, W.D., "The Physical Chemistry of Surface
    Films," pp. 209-210, Reinhold Publishing Corpora-
    tion, New York, N.Y., 1952.

20

Role of Double Interactions and Spreading Pressure in Particulate Soil Removal

HERMANN LANGE

Henkel and Cie., GmbH, Düsseldorf, Federal Republic Germany

Introduction

In the theory of particulate soil removal the influence of electric charges and double layers at the
surface of soil particles and substrates has been
discussed for many years (1-3). In analogy to the
Derjaguin-Landau-Verwey-Overbeek theory one takes into
account the van der Waals-London energy of attraction
P_A and the repulsive energy P_R due to electrical
double layer. By combining both types of interaction
energy one obtains diagrams of the type shown in
Figure 1. The curves have been calculated for two different values of the potential of the diffuse part of
the double layer using the tables given by Verwey and
Overbeek (4). The curve of the electrical repulsion
P_R as well as the maximum of the resultant curve is
shifted upwards when the potential rises. The potential may be increased, for example, by adsorption of
surface active ions.
This model has sometimes led to the conclusion
that not only the deposition of soil particles but also the removal of adhering particles will be inhibited
if the potential of the diffuse layers and, consequently, the maximum of the resultant curve is high.
It will be shown in this paper that this conclusion is
not justified because the choice of the common zero
point of the scale of potential energy is misleading.
Furthermore, the primary step of the removal of
adhering particles, namely the beginning of the penetration of an extremely thin layer of liquid between
particle and substrate allowing for the formation of
an adsorption layer will be discussed in the second
part.

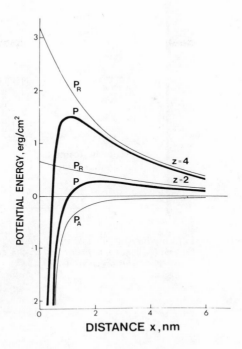

Figure 1.   Potential energies of attraction P_A and of repulsion P_R as function of the distance and the resultant P of them. $z =$ potential in units of $\psi_d \cdot e/(kT)$. $\psi_d =$ potential of the diffuse layer. The curves have been calculated for a Hamaker constant $A = 2 \cdot 10^{-13}$ erg and an electrolyte concentration of 10^{-2} mol/l (univalent ions).

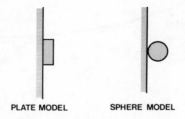

PLATE MODEL   SPHERE MODEL

Figure 2.   Two models for a particle adhering at a substrate

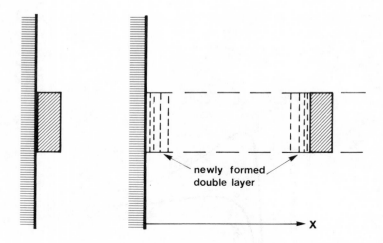

Figure 3. Removal of an adhering particle in an electrolyte solution. Plate model. Only the double layer newly formed in the contact zone after separation is shown.

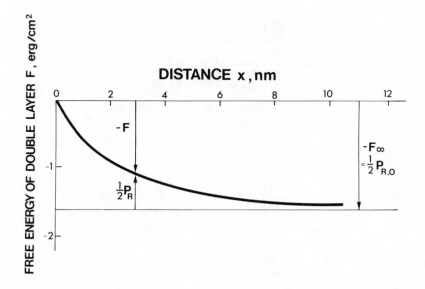

Figure 4. Free energy of the double layer per unit area as a function of the distance. Plate model.

Influence of the Electrical Double Layer on the Removal of Soil Particles

Two models of adhering soil particles will be taken into account as schematically shown in Figure 2. With the "plate model" the considerations are simpler than with the "sphere model". However, the sphere model gives a better description of most real soil particles.

Plate Model. The start point and the final state of the removal of a plate-like particle in an electrolyte solution is shown in Figure 3. In the beginning there will be an electrical double layer only at the outer surface of the system but not in the contact zone. After removal to a distance large enough that no interactions between particle and substrate are present new double layers are built up at the former contact area. The formation of the diffuse layers results in a decrease of the free energy of the system. At smaller distances an interaction of the double layers on both sides occurs. Consequently, the free energy is a function of the distance as shown in Figure 4. The decrease of free energy approaches asymptotically to a limiting value F_{∞} corresponding to the state of no interaction between the double layers. The amount of this value increases with rising potential of the diffuse layer.

In order to bring a particle from a large distance in contact with the substrate surface one has to do a work of $P = 2\,F_{\infty}$. The factor 2 is caused by the fact that the double layer is present at particle and substrate surface.

Hence, the value of F_{∞} is the correct reference level if one considers the approaching of a particle from a long distance to the substrate surface up to the contact. Consequently, the common zero level of the potential energy in the first diagram means two different values for F_{∞} for the lower ($z = 2$) and for the higher ($z = 4$) potential.

However, when considering the removal of adhering particles a zero level chosen in this way is no longer a correct common reference level for different potentials, because in the state of adhering no diffuse layer is present between the adherents. Consequently, the correct reference level for the removal of adhering particles is defined by the absence of any interaction energy between particle and substrate as well as of any free energy of double layer. So, for the potential energy one obtains curves as shown in Figure 5. $P_{A,0}$ is the free energy of adhesion caused by van der Waals-

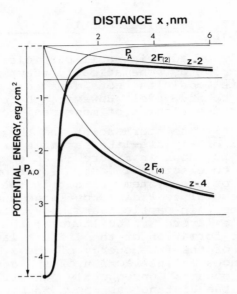

Figure 5.   Potential energy of attraction P_A
and free energy of double layer $2\,F_{(2)}$ and
$2\,F_{(4)}$ for the potentials $z = 2$ and $z = 4$.
The resultant curves are shown by the
solid lines. Plate model.

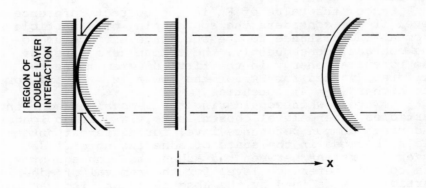

Figure 6.   Removal of an adhering particle in an electrolyte solution. Sphere
model.

London forces. The resultants of the free energies of
attraction and of the diffuse layers have been constructed.

From this diagram the important conclusion can be
drawn that with increasing potential the energy barriers not only for the deposition but also for the removal of particles will be lowered (5).

Sphere Model. The considerations with the sphere
model are somewhat more complicated than with the plate
model. Initially, there will be an interaction between
the diffuse layers of particle and substrate in a certain region around the point of contact as is to be
seen from Figure 6. Hence, the potential energy of the
system in the state of adhesion is no longer independent of the potential and other properties of the
double layer. The border of the region of double layer
interaction is arbitrary but it must be chosen to be
large enough to include all significant interaction.

Figure 7 shows the free energy of the diffuse
layer in the region of interaction. Indeed, the free
energy F_o in the region of interaction for the state
of adhesion is indefinite and depends on the choice of
the border of this region. However, the difference of
the free energies $P_{R,o}$ in the state of adhering and at
long distance is definite.

Although the term F_o is indefinite an arbitrary
alteration of it causes only a parallel shifting along
the ordinate axis. Therefore, it is usefull to chose
arbitrarily a value of zero for F_o. In this case, all
curves for different potentials and, consequently, for
different values of F_o have the same origin on the ordinate axis. At long distances, they approach asymptotically the level of $-P_{R,o}$ below the abscissa axis.

By combining with the curve for the van der Waals-
London attraction one obtains the diagram shown in
Figure 8. The result is, in principle, the same as that
one obtained with the plate model, namely, the height
of the energy barrier to be surmounted at the removal
of particles decreases with rising potential.

In all these considerations, however, it has been
assumed that the potentials at particle and substrate
are equal. As long as the difference between both
potentials is not too large the considerations will
remain valid in a qualitative sense. If anionic surfactants are adsorbed at the surfaces of particle they do
not only increase the negative potential but they will
also tend to equalize it to a certain degree.

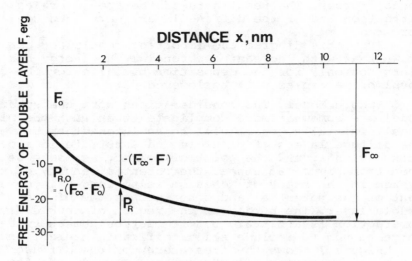

Figure 7. Free energy of the double layer in the region of interaction as a
function of the distance. Sphere model. $z = 4$.

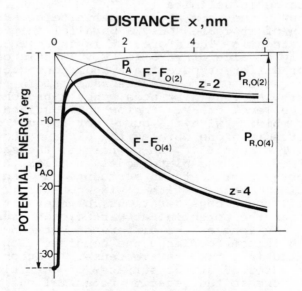

Figure 8. Potential energy of attraction P_A and the
difference $F-F_o$ as function of the distance, and the
resultants of them. Sphere model.

Some predictions following from the assumption
that electrical double layer forces are really effective in the washing process have been verified by
washing tests with homologous sodium alkyl sulfates
and variable concentration of added electrolyte (5-6).

The Primary Step of the Removal of Adhering Particles

The foregoing considerations account only for the
total height of the energy barrier to be surmounted at
the removal of adhering particles. However, special
attention must be paid to the beginning of the penetration of an extremely thin layer of liquid between particle and substrate allowing for the formation of an
adsorption layer.

The spreading pressure of the adsorbed layers of
surfactants exerts a disjoining force between particle
and substrate. This force favours the penetration of
liquid into the zone of contact. Figure 9 shows a model
of the system under consideration for a spherical particle.

The disjoining force may be calculated if the
values of the spreading pressures of the adsorbed layers at the surface of particle and substrate are known.
From Figure 9 the following equation can be derived by
simple geometrical considerations:

$$f_d = 2 \pi r \, (\pi_p + \pi_s)$$

f_d is the disjoining force, π_p and π_s are the spreading pressures of the adsorbed layers at particle and
substrate surface respectively. r is the particle radius.

The question arises wether this disjoining force
is sufficient for playing an essential part in the removal of adhering particles. For numerical calculations, values of the spreading pressures π_p and π_s
are necessary for the evaluation of the disjoining
force. Furthermore, experimental data on the force of
adhesion are required for comparison. It is somewhat
difficult to find out suitable data in the literature.

According to results obtained by Visser (7) the
average forces of adhesion f_s of carbon black particles of 0.1 μm radius on a cellulose film immersed in
water amount to approximately 23.10^{-6} dyn at pH 3.3
and 2.10^{-6} dyn at pH 10.

The spreading pressure of the adsorbed layer of
surfactants on hydrophobic particles in equilibrium
with the surfactant concentration c' may be evaluated
from adsorption isotherms by the formula (8)

$$\pi = RT \int_{0}^{c'} \frac{\Gamma(c)}{c} \, dc$$

where $\Gamma(c)$ is the adsorption as a function of the concentration c. For ionic surfactants, this formula is only valid when the counterion concentration is kept constant.

The adsorption isotherm shown in Figure 10 has been obtained for sodium dodecyl sulfate with added NaCl for constant sodium ion concentration at graphitized carbon black (Graphon, Cabot Corp. Boston, Mass.), The shape of the isotherm is very similar to that obtained by Day, Greenwood and Parfitt (9) for the same system without NaCl. By graphical integration of a plot Γ/c vs. c a value of π_p = 29 dyn/cm at c' = 8.10^{-3} mol/l (i.e. at the critical micelle concentration) has been obtained.

Alternatively, the spreading pressure of the adsorbed surfactant layer may be obtained from the difference between the wetting tensions for pure water and for the surfactant solution. A number of measurements of wetting tensions at paraffin wax with aqueous solutions of sodium alkyl sulfates (10) has been performed by the Guastalla method (11). Values of about 20 dyn/cm have been obtained at concentrations just below the critical micelle concentration.

Informations on the spreading pressure of the adsorbed layer on textile materials are hardly available. Some measurements of the wetting tension on different foils of textile-like materials have shown that the values are generally relatively low (10).

By inserting a rather low value for the sum of both spreading pressures, namely 20 dyn/cm into the equation one obtains for the radius 0.1 um

$$f_s = 125 \cdot 10^{-6} \text{ dyne}$$

This value is much higher than the range of the experimental values of the force of adhesion obtained by Visser. Hence, one may conclude from these estimations, that the disjoining force caused by the spreading pressure of the adsorbed layers will be sufficient to initiate the first step of the removal of adhering particles.

Summary

In the theory of the washing process, sometimes the influence of electric charges and their increase by

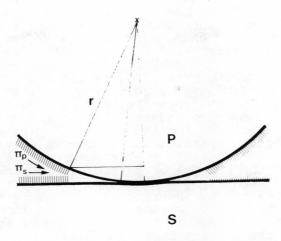

Figure 9.   Adsorption layers at particle P and substrate S, causing a disjoining force.

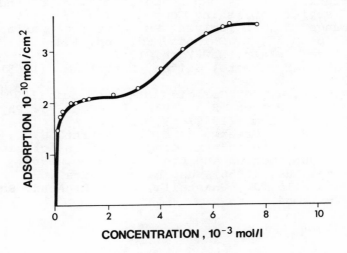

Figure 10.   Adsorption isotherm of sodium dodecyl sulfate (SDS) at Graphon with added sodium chloride. Sum of concentrations of SDS and sodium chloride constant and equal to 8.10^{-3} mol/l. Temperature $23 \pm 1°C$.

adsorption of ionic surfactants at the surfaces of
solid soil particles and substrate have been discussed. Whereas the inhibition of the deposition of particles by electric charges may be easily explained, a
satisfactory interpretation of the influence of the
charge on the elementary step of washing, i.e. the
removal of adhering particles, is lacking. This difficulty can be eliminated by choosing the correct reference states for the potential energies both for the
process of deposition and for that of removal. This
problem is discussed for the models of plate-like as
well as spherical particles. It follows that in both
cases an increase of the surface charge contributes
not only to the removal of particles. For the primary
step of the removal of adhering particles, the disjoining force caused by the spreading pressure of the
adsorbed layer of surfactants is also of importance.
An estimation shows that this force may be sufficient
to provoke the primary step of removal.

Literature Cited

1. Durham, K., J. Appl. Chem. (1956) $\underline{6}$, 153
2. Lange, H., Kolloid-Z. (1957) $\underline{154}$, 103
3. Lange, H., Kolloid-Z. (1958) $\underline{156}$, 108
4. Verwey, E.J.W., Overbeek, J.Th.G., "Theory of the
   Stability of Lyophobic Colloids", p. 82 and 141,
   Elsevier Publishing Company, New York, 1948
5. Lange, H., in "Solvent Properties of Surfactant
   Solutions", K. Shinoda, Ed., pp. 144-150,
   Marcel Dekker, New York, 1967
6. Lange, H., Fette, Seifen, Anstrichm. (1963) $\underline{65}$,
   231
7. Visser, J., J. Colloid Interface Sci.(1970)$\underline{34}$,26
8. Fu, Y., Hansen, R.S., Bartell, F.E., J. Phys. &
   Colloid Chem. (1949) $\underline{53}$, 1141
9. Day, R.E., Greenwood, F.G., Parfitt, G.D.,
   in Proc. 4th Intern. Congr. Surface Activity,Vol.2,
   p. 1005, Gordon and Breach, London, 1967.
10. Lange, H., (1956), unpublished data
11. Guastalla, J., Guastalla, L., C.R. Acad. Sci.(1948)
    $\underline{226}$, 2054

INDEX

INDEX